Plants of Central Asia

Volume 11

Plants of Central Asia

Plant Collections from China and Mongolia

(*Editor-in-Chief*: V.I. Grubov)

Volume 11

Amaranthaceae—Caryophyllaceae

Yu.D. Gusev
Yu.P. Kozhevnikov

CRC Press
Taylor & Francis Group
Boca Raton London New York

CRC Press is an imprint of the
Taylor & Francis Group, an **informa** business
A SCIENCE PUBLISHERS BOOK

First published 2007 by Science Publishers

Published 2019 CRC Press
Taylor & Francis Group
6000 Broken Sound Parkway NW, Suite 300
Boca Raton, FL 33487-2742

© 2007 by Taylor & Francis Group, LLC
CRC Press is an imprint of Taylor & Francis Group, an Informa business

No claim to original U.S. Government works

ISBN 13: 978-1-57808-123-3 (hbk) (vol 11)
ISBN 13: 978-1-57808-062-5 (set)

Visit the Taylor & Francis Web site at
http://www.taylorandfrancis.com

and the CRC Press Web site at
http://www.crcpress.com

Library of Congress Cataloging-in-Publication Data

Rasteniia Tsentral'noi Azii. English
  Plants of Central Asia: plant collections from   China   and
Mongolia
  / [editor-in-chief, V.I. Grubov].
      p. cm.
  Research based on the collections of the V.L. Komarov
Botanical Institute.
    Includes bibliographical references
  Contents: v. 11. Amaranthaceae—Caryophyllaceae
    ISBN 978-1-57808-123-3 (vol. 11)
      1. Botany--Asia, Central. I. Grubov, V.I. II.
Botanicheskii institut im. V.L. Komarova. III. Title.
QK374, R23613 2002
581.958--dc21   99-36729
                                                          CIP

Translation of : Rasteniya Tsentral' noi Azii, vol. 11, 1994
Nauka Publishers, Leningrad.

ANNOTATION

PLANTS OF CENTRAL ASIA. From the Material of the V.L. Komarov Botanical Institute, Russian Academy of Sciences, Vol. 11: Amaranthaceae —Caryophyllaceae. Compilers: Yu.D. Gusev and Yu.P. Kozhevnikov. 1994. "Mir i Sem'ya", St. Petersburg.

The eleventh volume of the illustrated lists of vascular plants of Central Asia (within the People's Republics of China and Mongolia) continues the description of flowering plants and covers families Amaranthaceae, Aizoaceae, Portulacaceae and Caryophyllaceae. Keys are provided for the identification of genera and species and references to nomenclature, and information on habitat and geographic distribution given for each species. The largest genera treated in this volume are *Stellaria* (30 species), *Silene* (28 species), *Arenaria* (27 species), *Gypsophila* (12 species) and *Cerastium* (11 species).

111.: 10 plates, 7 maps of distribution ranges.

V.I. Grubov
Editor-in-Chief

PREFACE

This volume concludes the description of the families of order Centrospermae (families 41-44), Caryophyllaceae being the main family among them. In Central Asia as a whole, it comprises 202 species distributed in 24 genera, the territory under consideration containing 148 species in 21 genera; the Central Asian part of the former USSR accounts for the rest of 54 species in 3 genera. Additionally, the list includes 12 species distributed in the adjoining regions (mainly Himalayan) and may perhaps be detected in Central Asia as well. The treatment of the family here does not recognise the independent status of such doubtful genera like *Dichodon*, *Elisanthe*, *Eremogone*, *Gastrolychnis*, *Mesostemma*, *Pleconax*, *Psammophiliella*, *Tyttostemma*, which have appeared in some recent floristic treatments. *Thylacospermum* has been cited among synonyms of genus *Arenaria* as done before by E. Fenzl, who originally established this genus.

At the same time, it does not appear to be correct to delete some species from the group *Arenaria capillaris* (*A. androsacea*, *A. formosa*), *Silene graminifolia* (*S. chamarensis*, *S. jenisseensis*, *S. iche-bogdo*) and some others which are quite distinct morphologically as well as ecologically and well distinguished in nature.

The largest genera of Caryophyllaceae are *Stellaria* (30 species), *Silene* (28 species), *Arenaria* (27 species), *Gypsophila* (12 species) and *Cerastium* (11 species). These 5 genera account for more than 2/3 of all species in the family, the rest of 15 genera containing 1 to 6 species each.

Endemism in the family for the territory under consideration is quite high with 24 endemic species, or 16% of their total number. Most of the endemic species belong to the largest genera: *Stellaria* (8 species), *Silene* (6) species, *Arenaria* (5 species). *Gymnocarpos przewalskii* Bge. ex Maxim. represents the most prominent endemic species. This is a small shrub with aciculiform leaves, forming nearly pure communities in solonetzic rubble-sand deserts of Mongolia. *G. fruticosus* Forsk., its close and lone relative, is distributed under similar arid conditions in the Mediterranean. Sometimes, this genus is treated as belonging to the family Illecebraceae. One way or the other, this morphologically distinct genus is of very ancient origin. Most of the endemic Caryophyllaceae species of Central Asia are characterised by distinctly manifest xeromorphic features: these are pulvinoid (*Stellaria pulvinata*) or compactly matted perennial plants

(*Gypsophila capituliflora* = *G. dschungarica*), or perennial plants with stout caudex (*Arenaria ischnophylla*). Very few among Central Asian Caryophyllaceae are, however, typical desert species. Apart from *Gymnocarpos przewalskii* mentioned above, *Gypsophila desertorum* can also be cited as a common member of desert and desert-steppe coenoses.

Most Caryophyllaceae are mountain-steppe and high-altitude species. Naturally, therefore, there are some 50 species in common with southern Siberia (Altay, Sayan mountains and Dauria) and 37 with Himalayas. The association with Himalayas is particularly noticeable in genus *Arenaria* which consists of 12 Tibet-Himalayan and Tien Shan-Tibet-Himalayan species (for example, *A. caespitosa*). On the other hand, the commonality of species with Middle Asia and Fore Asia as a whole is rather small at less than 30 species; this region has many annual plants among Caryophyllaceae, they being very rare in Central Asia. This commonality is far more distinct in Kazakhstan, western Central Asia. There are quite a large number of species in common with China (33 species) which, however, are restricted in their distribution mainly to Qinghai (for example, *Lepyrodiclis quadridentata, Arenaria kansuensis*).

On the whole, the composition of Caryophyllaceae in Central Asia is quite original, judging from the significant as well as ancient endemism, the group of characteristic life forms and geographical contiguities. Their significance in the vegetative cover of the region, however, is not as significant as that of Chenopodiaceae.

Apart from the desert species *Gymnocarpos przewalskii* and *Gypsophila desertorum* mentioned above, the widely distributed steppe species *Arenaria acicularis* and the high-altitude pulvinoid plants *Stellaria pulvinata, S. depressa* and several *Arenaria* species (*A. caespitosa, A. densissima, A. kansuensis, A. polytrichoides, A. pulvinata*) can be regarded as of coenotic importance.

The remaining 3 families treated in this volume, namely Amaranthaceae (7 species), Aizoaceae (1 species) and Portulacaceae (3 species) are extremely small, rare and do not play any significant role either in the flora or in the vegetative cover of Central Asia. These are essentially widely distributed weed and ruderal plants (*Amaranthus, Montia, Portulaca, Mollugo*) and arctic-mountain Asian species (*Claytonia*).

Thus, this volume covers a total of 171 species in 26 genera.

* * *

In this volume, O.I. Starikova translated the Chinese texts. O.V. Zaitseva prepared the drawings for plates. Yu.P. Kozhevnikov prepared the maps of distribution ranges. Family Amaranthaceae was processed by Yu.D. Gusev and families Aizoaceae, Portulacaceae, Caryophyllaceae by Yu.P. Kozhevnikov.

V.I. Grubov

CONTENTS

TAXONOMY

SPECIAL ABBREVIATIONS

Abbreviations of Names of Collectors

Bar.	—	V.I. Baranov
Chaney	—	R.W. Chaney
Ching	—	R.C. Ching
Czet.	—	S.S. Czetyrkin
Divn.	—	D.A. Divnogorskaya
Fet.	—	A.M. Fetisov
Grub.	—	V.I. Grubov
Gr.-Grzh.	—	G.E. Grum-Grzhimailo
Gub.	—	I.A. Gubanov
Ik.-Gal.	—	N.P. Ikonnikov-Galitzkij
Isach.	—	E.A. Isachenko
Kal.	—	A.V. Kalinina
Klem.	—	E.N. Klements
Kuan	—	K.C. Kuan
Lad.	—	V.F. Ladygin
Li	—	S.H. Li et al
Litw.	—	D.I. Litwinow
Lom.	—	A.M. Lomonossov
Mois.	—	V.S. Moiseenko
Merzb.	—	G. Merzbacher
Pal.	—	I.V. Palibin
Petr.	—	M.P. Petrov
Pias.	—	P. Ya. Piassezki
Pob.	—	E.G. Pobedimova
Pop.	—	M.G. Popov
Pot.	—	G.N. Potanin
Przew.	—	N.M. Przewalsky
Rachk.	—	E.I. Rachkovskaya
Reg. A.	—	A. Regel
Rob.	—	V.I. Roborowsky
Sap.	—	V.V. Sapozhnikov
Schischk.	—	B.K. Schischkin
Serp.	—	V.M. Serpukhov

2

Tug.   — A.Ya. Tugarinov
Yun.   — A.A. Yunatov
Zab.   — D.K. Zabolotnyi
Zam.   — B.M. Zamatkinov

Abbreviations of Names of Herbaria

B — Botanisches Museum, Berlin-Dahlem
BM — British Museum of Natural History, London
C — Botanical Museum and Herbarium, Copenhagen
Cal — Botanical Survey of India, Calcutta
G — Conservatoire et Jardin botaniques, Geneve (Geneva)
GH — Gray Herbarium of Harvard University, Cambridge, U.S.A.
GOET — Systematisch-Geobotanisches Institut, Universitat Gottingen
HK — Herbarium, Agriculture and Fisheries Department, Hong Kong
K — The Herbarium, Royal Botanic Gardens, Kew, Richmond, Surrey, London
KW — Botanical Institute of the Academy of Sciences of Ukraine, Kiev
LE — Herbarium of V.L. Komarov Botanical Institute of the Russian Academy of Sciences, St.-Petersburg
Linn. — Herbarium, The Linnean Society of London, London
MW — Biological Faculty of the State University, Moscow.
P — Museum National d'Histoire Naturelle, Paris
PE — Institute of Botany, Academia Sinica, Beijing
PR — Botanical Department of National Museum, Praha
TK — P.N. Krylov Herbarium of State University, Tomsk
TO — Instituto Botanico della Universita, Torino
U — Botanical Museum and Herbarium, Utrecht
UPS — Botanical Institute, University of Uppsala, Sweden
W — Naturhistorisches Museum, Wien (Vienna)
Z — Botanischer Garten und Institut fur Systematische Botanik der Universitat, Zurich

Family 41. **AMARANTHACEAE** Juss.

1. **Amaranthus** L.

Sp. pl. (1753) 989

1.   All inflorescences (small cymose racemes or glomerules) in leaf

axils and stem foliate up to tip .... 5.

\+ Small cymose racemes (glomerules) gathered in terminal, elongated, pseudospicate inflorescences, often paniculately branched and without leaves at least in upper part; quite often, reduced axillary inflorescences seen along stem and branches ....2.

2. Perianth segments and stamens 5 each. Fruit dehiscent on circular transverse fissure..3.

\+ Perianth segments and stamens 3 each. Fruit indehiscent...........4.

3. Plant pale-green, sometimes reddish. Perianth segments of pistillate flowers linear-cuneate, truncate at tip, somewhat longer than fruit. Inflorescence highly compact, usually with short stubby branches..2. **A. retroflexus** L.

\+ Plant dark-red or yellowish, rarely green. Perianth segments of pistillate flowers lanceolate, narrow-ovoid or narrow-elliptical, not longer than fruit. Inflorescence usually lax, often with long branches..1. **A. cruentus** L.

4 Leaves broad-emarginate or rounded at tip, abruptly cuneate-narrow at base. Besides elongated terminal false spikes, shortened axillary inflorescences present all along stem and branches. Bracts deltoid-lanceolate, 1/2 of perianth. Perianth segments narrow-lanceolate. Fruit 1.5-2.5 mm long, elliptical, glabrous or slightly rugose. Seed 1.2-1.5 mm in diam...................................7. **A. lividus** L.

\+ Leaves narrow-emarginate at tip, cordate, rounded or cuneate at base. Inflorescence terminal, paniculate, comprising elongated slender false spikes. Bracts ovoid to lanceolate, as long as perianth or shorter. Perianth segments of pistillate flowers narrow-obovoid; of staminate flowers oblong. Fruit 1-1.5 mm long, compressed-ovoid, strongly rugose. Seed 1-1.1 mm in diam.................................... ....6. **A. viridis** L.

5(1). Perianth segments 4, sometimes 5. Stamens 4. Stems usually procumbent, branched from base. Leaves obovoid, spatulate or oblong-spatulate. Seed 1.3-1.7 mm in diam......3. **A. blitoides** S. Wats.

\+ Perianth segments and stamens 3 each. Stems erect or ascending. Leaves ovoid and lanceolate to obovoid and spatulate....................6.

6. Bracts subulate, twice longer than perianth. Leaves oblong-obovoid or spatulate, more rarely elliptical, orbicular or slightly emarginate at tip, somewhat undulate on margin. Seed 0.8-1 mm in diam....... ....4. **A. albus** L.

\+ Bracts narrow- to broad-lanceolate, shorter than perianth. Leaves elliptical or rhombic-elliptical to ovoid and oblong. Seed 1.3-1.6 mm in diam...5. **A. sylvestris** Vill.

Section 1. **Amaranthus**

1. **A. cruentus** L. Syst. Nat. ed. 10, 2 (1759) 1269; Saner in Ann. Missouri Bot. Gard. 54, 2 (1967) 122; Hanelt in Kultur pflanze, 16 (1968) 129; Gusev in Bot. zh. 57, 5 (1972) 460. —*A. paniculatus* L. Sp. pl. ed. 2 (1763) 1406; Forbes and Hemsley, Index Fl. Sin. 2 (1902) 320; Vassilczenko in Fl. SSSR, 6 (1936) 361; Fl. Kirgiz. 5 (1955) 84; Fl. Kazakhst. 3 (1960) 324; Fl. Tadzh. 3 (1968) 462; Fu Hiang-chian in Fl. Intramong. 2 (1978) 151; Kuan Ke-chien in Fl. Sin. 25, 2 (1979) 207. —**Ic.**: Hegi, Ill. Fl. Mitt. Eur. 2 Aufl., B. III/2, Lief. 1 (1959) Fig. 208, 239 f-h.

Described from America. Type in London (Linn.).

In oases, sometimes as weed in kitchen gardens; among crops and as wild growth.

IA. Mongolia: *Alashan* (Dyn-Yuan—in oasis, in kitchen gardens, among crops, June 5-6, 1908—Czet.).

General distribution: Cent. Tien Shan; Europe (except north), Caucasus, Middle Asia, East. Sib. (south), Far East (Primor'e), Nor. Mong. (Ulan-Bator), China (nearly entirely), Japan, Indo-Mal., Nor. Amer., Africa, Austral.

Note. In warm and warm-temperate regions of the world, cultivated as ornamental plant; rarely as cereal plant. Sometimes as weed. Originally from Cent. Amer.

2. **A. retroflexus** L. Sp. pl. (1753) 991; Forbes and Hemsley, Index Fl. Sin. 2 (1902) 320; Kryl., Fl. Zap. Sib. 4 (1930) 977; Vassilczenko in Fl. SSSR, 6 (1936) 362; Kitag. Lin. Fl. Mansh. (1939) 193; Fl. Kirgiz. 5 (1955) 84; Grubov, Konsp. fl. MNR [Conspectus of Flora of Mongolian People's Republic] (1955) 125; id. in Bot. mat. (Leningrad) 19 (1959) 537; Fl. Kazakhst. 3 (1960) 323; Fl. Tadzh. 3 (1968) 463; Gusev in Bot. zh. 57, 5 (1972) 460; Fu Hiang-chian in Fl. Intramong. 2 (1978) 148; Kuan Ke-chien in Fl. Sin. 25, 2(1979) 208; Grubov, Opred. rast. Mong. [Key to Plants of Mongolia] (1982) 98. —**Ic.**: Fl. Sin. 25, 2, tab. 46, fig. 4-6.

Described from Nor. Amer. (Pennsylvania). Type in London (Linn.).

As weed in kitchen gardens and flowerbeds.

IA. Mongolia: *East. Mong.* (Choibalsan town, bank of Kherulen river, weed in Chinese kitchen gardens, Aug. 2, 1949—Yun.), *East. Gobi* (Dalan-Dzadagad town, left bank of brook irrigating a nursery of fodder grasses, weed, July 28, 1951—Kal.; sand along bank of Buir-Nur lake, near Khalkhin-Gol river estuary, Sept. 9, 1980—Gub.).

IIA. Junggar: *Jung.-Ala Tau* (Dzhair mountain range, Tushandzy settlement, alongside of irrigated flowerbed, Aug. 1951—Mois.).

General distribution: Aralo-Casp., Fore Balkh., Jung.-Tarb., Nor. Tien Shan, Cent. Tien Shan; Europe, Mediterr., Balk.-Asia Minor, Fore Asia, Caucasus, Mid. Asia, West. Sib., East. Sib. (south), Far East (south), Nor. Mong., China, Korean peninsula, Nor. Amer., South Amer. (south), Afr., Austral., N. Zealand.

Note. Originally from Nor. Amer. Naturalized in almost all, except cold, countries. where it is quite often an introduced plant.

Section 2. **Blitopsis** Dumort

3. **A. blitoides** S. Wats. in Proc. Amer. Acad. Arts Sci. 12 (1877) 273; Vassilczenko in Fl. SSSR, 6 (1936) 363; Kitag. Lin. Fl.. Mansh. (1939) 193; Fl. Kirgiz. 5 (1955) 84; Fl. Kazakhst. 3 (I960) 323; Fl. Tadzh. 3 (1968) 463; Gusev in Bot. zh. 57, 5 (1972) 461; Fu Hiang-chian in Fl. Intramong. 2 (1978) 148; Kuan.Ke-chien in Fl. Sin. 25, 2 (1979) 214; Gubanov in Byul. Mosk. o-va ispyt. prir. 87, 1 (1982) 126; Grubov, Opred. Rast. Mong. [Key to Plants of Mongolia] (1982) 269. —Ic.: Fl. SSSR, 6, Plate 18, fig. 7.

Described from western Nor. Amer. Type in Cambridge (GH).

IA. **Mongolia:** *East. Mong.* (40 km east of Choibalsan town, on garbage in Sumber state farm settlement, Sept. 12, 1980—Gub.).

General distribution: Aralo-Casp., Cent. Tien Shan; Europe, Mediterr., Caucasus, Mid. Asia, West. Sib., East. Sib. (south), Far East (south), China (Dunbei, North, South), Nor. Amer.

Note. Originally from Nor. Amer.

4. **A. albus** L. Syst. Nat., ed. 10, 2 (1759) 1268; Kryl. Fl. Zap. Sib. 4 (1930) 978; Vassilczenko in Fl. SSSR, 6 (1936) 364; Fl. Kirgiz. 5 (1955) 85; Fl. Kazakhst. 3 (1960) 321; Fl. Tadzh. 3 (1968) 465; Gusev in Bot. zh. 57, 5 (1972) 461; Kuan Ke-chien in Fl. Sin. 25, 2 (1979) 213. —Ic.: Fl. SSSR, 6, Plate 18, fig. 6.

Described from Nor. Amer. (Philadelphia). Type in London (Linn.).

On garbage around houses.

IIA. Indicated for Sinkiang without precise location of find (Kuan Ke-chien, l.c.).

General distribution: Aralo-Casp., Fore Balkh., Nor. Tien Shan, Cent. Tien Shan; Europe, Mediterr., Balk.-Asia Minor, Fore Asia, Caucasus, Mid. Asia, West. Sib., East. Sib. (south), Far East (south), Nor. Mong. (Hent.), China (Dunbei, North), Nor. Amer., Afr., Austral.

Note. Originally from Nor. Amer.

5. **A. sylvestris** Vill. Cat. pl. Jard. Strasb. (1807) 111. —*A. graecizans* L. subsp. *sylvestris* (Vill.) Brenan in Watsonia, 4, 6 (1961) 273; Gusev in Bot. zh. 57, 5 (1972) 462. —*A. roxburghianus* Kung in Bot. Jour. Fl. Nor. Sin. 4 (1935) 19, tab. 8; Kuan Ke-chien in Fl. Sin. 25, 2 (1979) 214. —*A. blitum* auct. non L.: Vassilczenko in Fl. SSSR, 6 (1936) 364; Fl. Tadzh. 3 (1968) 464. —Ic.: Fl. SSSR., 6, Plate 18, fig. 8 (sub nom. *A. blitum*).

Described from West. Europe. Type in Paris (P).

In fields and on garbage.

IA. **Mongolia:** *Alash. Gobi* ("Ninsya"—Kuan Ke-chien, l.c.).

IB. **Kashgar:** *Nor.* (between Kucha town and Kurlya settlement, near Bugur village, in fields, Aug. 20, 1929—Pop.), *East.* (8 versts—1 verst = 1.067 km—from Turfan, Chitkal' area, Sept. 14, 1898—Klem.).

IIA. Junggar: *Dzhark*. (Kul'dzha, May 1877—A. Reg.).
General distribution: Europe, Mediterr., Fore Asia, Caucasus, Mid. Asia, China (North, North-West, Cent. South), Himalayas (Kashmir), Indo-Malay, Afr., Austral.

6. A. viridis L. Sp. pl., ed. 2 (1763) 1405; Gusev in Bot. zh. 56, 9 (1971) 1359; id. in Bot. zh. 57, 5 (1972) 462; Kuan Ke-chien in Fl. Sin. 25, 2 (1979) 216. —Ic.: Fl. Sin. 25, 2, tab. 47, fig. 5-8.
Described from Europe and Brazil. Type in London (Linn.).
As weed in fields, kitchen gardens and flowerbeds.

IB. Kashgar: *South*. (Khotan district, Azna-bazar, weed, Oct. 25, 1940—T.T. Trofimov).
General distribution: Mid. Asia (south), Mediterr., Caucasus (Azerbaidzhan), China, Himalayas, Korean peninsula (south), Japan, Indo-Malay, Amer., Afr., Austral.

Note. As weed plant and in crops (vegetable). Originally, probably from tropical America.

7. A. lividus L. Sp. pl. (1753) 990; Vassilczenko in Fl. SSSR, 6 (1936) 366; Fl. Tadzh. 3 (1968) 466; Hanelt in Kultur pflanze, 16 (1968) 139; Gusev in Bot. zh. 57, 5 (1972) 463; Kuan Ke-chien in Fl. Sin. 25, 2 (1979) 217. —"*A. blitum*" L. Sp. pl. (1753) 990, nom. ambig.; Forbes and Hemsley, Index Fl. Sin. 2 (1902) 319, p.p.; Fl. Kazakhst. 3 (1960) 321. —*A. angustifolius* auct. non Lam.: Kryl. Fl. Zap. Sib. 4 (1930) 979. —Ic.: Fl. SSSR, 6, Plate 18, fig. 10.
Described from cultivated plants of the 18th century. Neotype in London (BM).
In kitchen gardens, gardens, roadsides, and river banks as weed and ruderal plant.

IA. Mongolia: *Khesi* (left bank of Huang He river valley north of Lanzhou town, July 10, 1875—Pias.).
IB. Kashgar: *East*. (Nor. fringe of Khami desert, ploughed fields, Aug. 18, 1895— Rob.).
General distribution: Fore Balkh., Cent. Tien Shan; Europe (except north), Mediterr., Balk.-Asia Minor, Fore Asia, Caucasus, Mid. Asia, West. Sib. (Altay), Far East, China, Himalayas, Korean peninsula, Japan, Indo-Mal., Nor. Amer., Afr., Austral., N. Zealand.

Note. Probably, originally from Europe.

Family 42. AIZOACEAE Rud.

1. Mollugo L.

Sp. pl. (1753) 89

1. M. cerviana (L.) Ser. in DC. Prodr. 1 (1824) 391; Fenzl in Ann. Wien. Mus. 1 (1836) 375; id. Monogr. Mollug. (1839) 261; Boiss. Fl. or. (1867) 756; Kryl. Fl. Zap. Sib. 5 (1931) 981; Kuzen. in Fl. SSSR, 6 (1936) 374; Grub.

Konsp. fl. MNR [Conspectus of Flora of Mongolian People's Republic] (1955) 125; id. Opred. rast. Mong. [Key to Plants of Mongolia] (1982) 98. — *Pharnaceum cerviana* L. Sp. pl. (1753) 272. —*P. umbellatum* Forsk. Fl. Aegypt.- arab. (1775) 58. —*Mollugo umbellata* Ser. in DC. Prodr. 1 (1824) 393. —**Ic.:** Fl. SSSR, 6, Plate 19, fig. 3; Grub. Opred. rast. Mong. [Key to Plants of Mongolia] Plate 41, fig. 194.

Described from Europe. Type in London (Linn.).

In sandy deserts, more often as weed around grazing grounds and nomadic camps.

IA. Mongolia: East. *Gobi* (Shine-Usu khuduk area on Sain-Shanda—Ulegei-khid telegraphic line, fine hummocky sand covered with sparse saxaul, Sept. 19, 1940—Yun.), **West. *Gobi*** (Shara-Khulusun area, on gorge floor, Aug. 8, 1943—Yun.; Ederingiin-Nuru, southern slopes, 75 km nor. of Shara-Khulusuin bulak spring, ephedra desert, Aug. 27, 1973—Isach. and Rachk), **Trans-Altay *Gobi*** (east of Khutsyn-Shand area, mud volcanos around spring, ravine valley, sand-covered rocks on rocky lowland, Sept. 2, 1979—Grub., Dariima et al).

General distribution: Fore Balkh. Europe, Mediterr., Balk.-Asia Minor, Afr., Austral.

Family 43. PORTULACACEAE Juss.

1. Flowers yellow, sessile, with deciduous sepals. Capsule dehiscent on operculum. Annual plants ......1. **Portulaca** L. (*P. oleracea* L.).
+ Flowers white or pink, on pedicels, sepals not deciduous. Capsule dehiscent on 3 valves. .... 2.
2. Perennial, 10-22 cm tall plants with rosette of radical oval leaves. Stamens 5. Petals 8-12 mm long, white or pink. ....
.... 2. **Claytonia** L. (*C. joanneana* Roem. et Schult.).
+ Small, 2-6 cm tall annual plants without rosette of radical leaves. Stamens 3. Petals 2-3 mm long, white...3. **Montia** L. (*M. fontana* L.).

1. Portulaca L.

Sp. pl. (1753) 445

1. **P. oleracea** L. l.c.; DC. Pr. 3 (1828) 353; Bunge, Enum. pl. China bor. (1832) 30; Fenzl in Ledeb. Fl. Ross. 2 (1842) 145; Boiss. Fl. or. 1 (1867) 757; Forbes and Hemsl. in J. Linn. Soc. Bot. 23 (1886) 71; Kuzen. in Fl. SSSR, 6 (1936) 386; Grub. Konsp. fl. MNR [Conspectus of Flora of People's Republic of Mongolia] (1955) 125; id. Opred. rast. Mong. [Key to Plants of Mongolia] (1982) 98; Ma Yu-chuan, Wang Zhi-min in Fl. Intramong. 2 (1978) 154; An Zheng-Xi in Claves. pl. Xinjiang. 2 (1983) 232; Fl. Sin. 1 (1985) 44. —**Ic.:** Hegi, Ill. Fl. Mitt. Eur. 3 (1912) tab. 98, fig. 2, 563; Fl. Intramong. 2, tab. 84; Fl. Sin. 1, tab. 105, fig. 7-11.

Described from Europe. Type in London (Linn.).

On banks of rivers and lakes, more often as weed in villages and on cultivated land.

IA. **Mongolia:** *Khesi* (nor. portion of Gansu province, Tash'-zhe, Mar. 27, 1909—Dimidenko), *Alash.* **Gobi** (Dyn'yuanin oasis, as weed among vegetable crops, June 6, 1908—Czet.).

IB. **Kashgar:** *West.* (Yarkand oasis, 900 m, on moist loessial clay, June 5, 1989—Rob.; Yangi-Gissar region, Kizyl' village, in maize field, Aug. 13, 1909—Divn.; Kashgar oasis, Yangi-Abad village, in farms, July 31, 1929—Pop.), *East.* (Khami desert, Changlyufi basin, kitchen garden, Aug. 18, 1895—Rob.).

IIA. **Junggar:** *Dzhark.* (Kul'dzha, July 1877; Suidun, July 16, 1877—A. Reg.; Chitkal' area, Sept. 14, 1898—Klem.).

General distribution: Europe, Fore Asia, Caucasus, Far East, Nor. Mongolia (Hent., Hang.), China (North, North-West, Cent.), Japan.

Note. In Kunlun mountains, *Portulaca grandiflora* Hook. is cultivated in gardens and on house roofs.

2. Claytonia L.

Sp. pl. (1753) 204

1. **C. joanneana** Roem. et Schult. Syst. nat. 5 (1819) 434; Maxim. Enum. pl. Mong. (1889) 106; Kryl. Fl. Zap. Sib. 5 (1931) 982; Kuzen. in Fl. SSSR, 6 (1936) 381; Grub. Konsp. fl. MNR [Conspectus of Flora of People's Republic of Mongolia] (1955) 125; id. Opred. rast. Mong. [Key to Plants of Mongolia] (1982) 98. —*C. acutifolia* Ledeb. Fl. Alt. 1 (1829) 253. —*C. arctica* var. *joanneana* (Roem. et Schult.) Trautv. in Acta Horti Petrop. 5, 1 (1877) 56. —Ic.: Grub. Opred rast. Mong. [Key to Plants of Mongolia] Table 41, fig. 191.

Described from Siberia. Type in London (Linn.).

In meadows and marshes, on banks of brooks and rivulets, on arid slopes of mountains in alpine belt and in upper part of forest belt.

IA. **Mongolia:** *Khobd.* (around Ubsa lake, July 13, 1879—Pot.; Turgen' mountain range, Turgen'-gola valley 7 km above estuary, *Cobresia* mixed grass field on nor. slope, Aug. 17. 1971—Grub., Ulzij. et al), *Mong. Alt.* (upper Tsagan-gola, rocky ridge on Ozernoe glacier, on placers, July 6, 1905—Sap.).

General distribution: Jung.-Tarb.; Arct. (Asian), West. Sib. (Alt.), Nor. Mong. (Fore Hubs., Hent., Hang., Mong.-Daur.).

3. Montia L.

Sp. pl. (1753) 87

1. **M. fontana** L. l.c.; Henderson and Hume, Lahore to Yarkand (1873) 312; Walters in Watsonia, 3, 1 (1953) 4; Kozhevn. in Bot. zh. 63, 7 (1978) 697; Jage in Hegi, Ill. Fl. Mitt. Eur., 3, 2 (1979) 1214. —*M. lamprosperma* Cham. in Linnaea, 6 (1831) 564; Kuzen. in Fl. SSSR, 6 (1936) 385. —*M. rivularis* C.C. Gmel. Fl. Baden. (1806) 302; Kuzen. in Fl. SSSR, 6 (1936) 385. —Ic.: Fl. SSSR, 6, Plate 19, fig. 7.

Described from Europe. Type in London (Linn.).
In humid lowlands around brooks.

IIIC. Pamir (Tagdumbash-Pamir, at about 4500 m alt.—Deasy, 1901).
General distribution: Arct. (Europ.), Europe, Mediterr., Balk.-Asia Minor, Fore Asia, Caucasus, East Sib., Far East, Japan, Nor. Amer., South Amer., Afr., Austral., N. Zealand.

Family 44. CARYOPHYLLACEAE Juss.

1. Leaves with scarious stipules..2.
+ Leaves without stipules...4.
2. Shrubs....12. **Gymnocarpos** Forsk. (*G. przewalskii* Bunge ex Maxim.).
+ Grasses...3.
3. Very small flowers (about 1 mm in diam.) gathered in glomerules in axils of upper leaves or spike-like. Capsule scarious, 1-seeded. ....13. **Herniaria** L.
+ Flowers longer (calyx 3 mm or longer), in loose racemes. Capsule coriaceous, many-seeded..............11. **Spergularia** (Pers.) J. et C Presl.
4. Sepals free or adnate only at base, but not forming a tube..............5.
+ Sepals adnate into a more or less inflated tube or up to their middle ....14.
5. Petals emarginate at tip or laciniated almost down to base...........6.
+ Petals entire, rounded at tip, truncate or somewhat emarginate....8.
6. Notch on petals reaches 1/3 or less of their length ....4. **Cerastium** L.
+ Notch on petals reaches 1/2 or more of their length (sometimes petals reduced)...7.
7 Sepals acute. Capsule dehiscent on 6 (rarely 4) entire valves ....1. **Stellaria** L.
+ Sepals globose. Capsule dehiscent on 5 valves, bidentate at tip.. .... 3. **Myosoton** Moench.
8. Long-stalked cleistogamous flowers present in lower part of stem apart from ordinary flowers. Roots tubercular .... 2. **Pseudostellaria** Pax ex Pax et Hoffm.
+ Only ordinary flowers (chasmogamous) present in upper portion of shoots. Roots not tubercular ....9.
9. Plants very small, with axillary peduncles almost from base of stem ....7. **Sagina** L.
+ Plants relatively large (over 5 cm tall), bed- or cushion-forming...10.
10. Capsule dehiscent on 3 entire valves. ....8. **Minuartia** L.
+ Number of valves on capsule different ....11.
11. Capsule dehiscent on 2 valves. Styles 2 (rarely 3). Large (up to 0.5 mm tall) annual plants with broad leaves......5. **Lepyrodiclis** Fenzl.

+ Capsule dehiscent on 6 valves or teeth. Styles 3. Perennial plants (if annual, not taller than 12-15 cm)... 12.

12. Inflorescence umbelliform. Seeds peltate. Annual plants ....
....6. **Holosteum** L.

+ Inflorescence not umbelliform; seeds not peltate. Mainly perennial plants ....13.

13. Valves on capsule convoluted outwardly. Seeds glabrous, lustrous, with protuberance at point of attachment. Leaves oblong-oval or broad-lanceolate ....10. **Moehringia** L.

+ Capsule dehiscent on 6 teeth. Seeds tubercular, matte, without protuberance at point of attachment. Leaves setaceous, orbicular or ovoid ....9. **Arenaria** L.

14. Calyx with commissural nerves. Styles 3 or 5 ....15.

+ Calyx without commissural nerves. Styles 2 ....17.

15. Styles 3. Capsule dehiscent on 6 teeth ....14. **Silene** L.

+ Styles 5. Capsule dehiscent on 5 or 10 teeth ....16.

16. Calyx urceolate. Capsule dehiscent on 10 teeth ....
....16. **Melandrium** Roehl.

+ Calyx obconical or clavate. Capsule dehiscent on 5 teeth.............
....15. **Lychnis** L.

17. Calyx strongly inflated at base and narrowed (during anthesis) at tip, with oblong winglike projections ......20. **Vaccaria** N.M. Wolf.

+ Calyx not inflated, without winglike projections ....18.

18. Capsule scarious in lower portion, dehiscent there on transverse fissure. Leaves thorny. Stems many, forming thorny globose cluster .... 19. **Acanthophyllum** C.A. Mey.

+ Capsule coriaceous, dehiscent on 4 teeth. Leaves not thorny. Stems few; if many, not forming globose cluster ....19.

19. Calyx tubular, with 1-4 pairs of glumes ....21. **Dianthus** L.

+ Calyx campanulate, without glumes......................................20.

20. Sepal nerves thickened, calyx thus appearing ribbed. Seeds peltate..
.... 18. **Petrorhagia** (Ser. in DC) Link.

+ Sepal nerves not thickened. Seeds reniform ... ..17. **Gypsophila** L.

1. **Stellaria** L.

Sp. pl. (1753) 421

1. Plants with cushion-forming type of growth (proximated short stems lignifying in lower portion). Peduncles barely projecting above cushion surface (flowers as though resting on cushion).......2.

+ Plants not pulvinoid (stems of varying length, herbaceous down to base). Peduncles generally projecting above leaf-covered portion of plant ....3.

2. Stems glabrous. Base of some leaves with only stray cilia. Cushions highly compact, humped at middle. Petals absent. Calyx invariably 5-merous. Styles invariably 3 ....24. **S. pulvinata** Grub.

+ Stems sparsely pubescent with crispate hairs. Leaves with many cilia along margin. Cushions rather loose, flat. Petals lacinated into linear lobes, 1/3 shorter than sepals. Calyx 4-5-merous. Styles 2-3.. ....19. **S. maximowiczii** Ju. Kozhevn.

3. Inflorescence umbellate.. .... 18. **S. irrigua** Bge.

+ Inflorescence not umbellate.. .... 4.

4. Flowers 4-merous (4 each sepals and petals). Stamens 8. Styles 2. Capsule dehiscent on 4 valves.. .... 5.

+ Flowers of mixed structure (perianth 5-merous). Stamens 8-10. Styles 2-3 (rarely 4). Capsule dehiscent on 4-5-6 valves ....7.

5. Entirely glabrous annual plants with relatively linear leaves. Petals laciniated to 1/4 or less........ ....2. **S. alsinoides** Boiss. et Buhse.

+ Perennial plants with lanceolate leaves. Stems, leaves and sepals more or less pubescent. Petals laciniated up to middle............ ..... 6.

6. Plants glandular-pubescent or glabrous. Petals 1.5 times longer than sepals, pubescent with glandular hairs at back ....
.... **S. schugnanica** Schischk.

+ Plants with simple curved hairs. Petals shorter than or equal to sepals, with simple hairs on back..1. **S. alexeenkoana** Schischk.

7. Plants entirely glabrous ....8.

+ Plants fairly pubescent (sometimes with cilia only at leaf base) ..
.... 11.

8. Calyx (2.5) 3-4 (4.5) mm long. Bracts herbaceous, foliaceous (peduncles emerge from axils of upper leaves). Plants generally up to 20 cm tall... ..9.

+ Calyx 4-7 (7.5) mm long. Bracts scarious. Plants 20-30 cm tall....10.

9. Stems weak, lodging, 8-15 (20) cm long. Flowers in terminal inflorescences or single. Sepals green, long-cuspidate, with 3 distinct nerves. Capsule only slightly longer than calyx. Leaves lanceolate, long-cuspidate .... .8. **S. crassifolia** Ehrh.

+ Stems strong, erect, 2-5 cm tall, frequently forming small beds. Flowers single. Sepals purple-colored, short-cuspidate, with 1 distinct nerve. Capsule 1.5-2 times longer than calyx. Leaves orbicular-ovoid, short-cuspidate...30. **S. winkleri** (Briq.) Schischk

10. Stems branched from base. Leaves 0.7-2 cm long. Petals nearly equal to sepals, 4-5 mm long ....13. **S. dilleniana** Moench.

+ Stems usually simple (if branched from base, plant more than 30 cm tall). Leaves 3-5 cm long, standing almost perpendicular from stem. Petals up to twice longer than sepals, 5-7 mm long ....
....22. **S. palustris** Retz.

11. Stems dichotomously branched many times, easily breaking at nodes, forming subglobose cluster. Root stout (about 5 mm in diam.) entering soil vertically deeply... ..12.

\+ Stems not dichotomously branched, not breaking at nodes. Roots not thick (up to 2 mm in diam.), entering soil angularly ....15.

12. Plants without glandular pubescence ....13.

\+ Glandular pubescence on different plant parts, especially in upper portion of shoots ....14.

13. Sepals subobtuse, 3.5-4.5 (5) mm long. Capsule 1-seeded, invariably 6-valved. Styles invariably 3..............17. **S. gypsophylloides** Fenzl.

\+ Sepals sharp, 4.5-5.5 (6) mm long. Capsule 2-3-seeded, 4-6-valved. Styles 2-3 ....16. **S. gyangtsensis** Will.

14. Sepals sharp, with narrow scarious margin, 1-1.5 mm broad. Petals laciniated into narrow lobes ....12. **S. dichotoma** L.

\+ Sepals short-cuspidate or blunt, with broad scarious margin, 2-2.5 mm broad. Petals laciniated into broad lobes............................
....3. **S. amblyosepala** Schrenk.

15. Pubescence of simple and glandular or only glandular hairs...16.

\+ Pubescence of only simple curved hairs, sometimes only cilia....21.

16. Leaves sessile, oval-oblong or linear ....17.

\+ Leaves (often only lower) petiolate, cordate or oval, long-cuspidate
.... 19.

17. Calyx 9 mm in diam. Petals 1/3 shorter than sepals. Styles 4. Capsule dehiscent on 4 valves...28. **S. strongylosepala** Hand.-Mazz.

\+ Calyx less than 8 mm in diam. Petals shorter, equal to or longer than sepals. Styles 3 (2). Capsule dehiscent on 5-6 valves .... ..18.

18. Capsule not inflated, dehiscent on valves or semivalves. Plant up to 25 cm tall, with 3-4 cm long leaves .... **S. tibetica** Kurz.

\+ Capsule inflated, dehiscent on teeth, reflexed outwardly. Plant up to 10 cm tall, with 1-2 cm long leaves .... **S. dianthifolia** Will.

19. Glandular pubescence only on sepals (sometimes pubescence consists of only simple hairs). Leaves less than 1.5 cm long and 1 cm broad, short-cuspidate. Petals shorter than sepals or absent
....20. **S. media** (L.) Cyr.

\+ Glandular pubescence on all parts of plants. Leaves longer than 1.5 cm and 1 cm broad, long-cuspidate. Petals longer than or equal to sepals, rarely shorter... ..20.

20. Capsule unilocular. Sepals ovoid, blunt, 4-6 mm long. Stem cross-section orbicular. Leaves ovoid. Seeds echinate-tuberculate ....
....6. **S. bungeana** Fenzl.

\+ Capsule 3-locular. Sepals oblong-lanceolate, acute, 3-4 mm long. Stems tetragonal. Leaves broad-lanceolate. Seeds rugose.. ....
.... **S. monosperma** Buch.-Ham.

21. At least some parts of plant compactly pubescent (green surface not visible) .... 22.
+ All parts of plant sparsely or diffusely pubescent (green surface distinctly visible) .... 24.
22. Petals fimbriate on upper margin. Stems tall, erect, emerging singly from rhizome (not aggregated) ....26. **S. radians** L.
+ Petals bifid. Stems ascending or creeping (plaited), emerging together from rhizome .... 23.
23. Leaves oval or broad-ovate, sparsely pubescent above and beneath. Flowers 5-10 mm in diam. Petals 1.5 times longer than subobtuse or short-cuspidate sepals. Capsule shorter than calyx.. ....
.... **S. turkestanica** Schischk.
+ Leaves lanceolate or broad-lanceolate, glabrous above and white tomentose pubescence beneath. Flowers about 3 mm in diam. Petals much shorter than lanceolate, long-cuspidate sepals (barely visible). Capsule twice longer than calyx .... **S. lanata** Hook. f.
24. Stems strong, erect (if elongated), forming compact or loose mat. Leaves thickened along margins, compactly imbricated or forming fascicles at nodes of elongated stems..
....7. **S. cherleriae** (Fisch.) Will.
+ Stems weak, often lodging, forming loose bed. Leaves not thickened along margin, not imbricated, evenly distributed on stem .... 25.
25. Plant with long rhizome, with fascicles of rather thick ascending stems, not more than 10 cm long emerging from rhizome. Flowers with 5-lobed nectar disk. Capsule 3-5 valved. Perianth 4-5-merous..
.... 26.
+ Plant with short rhizome, with solitary slender stem usually more than 10 cm tall emerging from rhizome. Nectar disk absent. Capsule 4-6-valved. Perianth invariably 5-merous .... 29.
26. Leaves rigid, lustrous, ovoid, long-cuspidate, carinate, profusely ciliate along margins. Styles invariably 3 ..4. **S. arenaria** Maxim.
+ Leaves soft, mat, lanceolate or oblong-oval, short-cuspidate, sparsely ciliate, mainly at base. Styles 2-3 (4) .... 27.
27. Flowers in terminal inflorescences with several pairs of large bracts, entirely scarious or with green midnerve. Peduncles glabrous. Sepals 3-5 (7)-nerved (var. *arenicola* with indistinct nerves on sepals). Styles 3 .... 11. **S. depressa** Schmid.
+ Flowers single, without bracts. Peduncles with sparse hairs. Sepals with 1, sometimes 3, distinct nerves. Styles 2 .... 28.
28. Leaves 8-12 mm long, oblong-elliptical, mat. Beds very loose, mainly with lodging stems. Petals equal to or slightly longer than sepals. Seeds punctate-tuberculate ....
....14. **S. divnogorskaje** Ju. Kozhevn.

+   Leaves about 4 mm long, lanceolate, lustrous. Beds compact, mainly with erect stems. Petals very small. Seeds subglabrous.. . ...10. **S. decumbens** Edgew.

29.   Leaves 3-6 cm long, linear, often standing perpendicular to stem. Inflorescences few-flowered, with 1-2 pairs of bracts; bracts scarious or leafy with scarious margins; terminal flowers axillary, on long slender peduncles longer than leaves.. .... 27. **S. soongorica** Roshev.

+   Leaves 0.7-4 cm long, lanceolate, not deflexed perpendicular to stem. Flowers in many-flowered infloresces with several pairs of scarious bracts.. .... 30.

30.   Mature capsules turning brown or black.. .... 34.

+   Mature capsules green or straw-yellow.. .... 31.

31.   Sepals ovoid, with broad scarious margin, short-cuspidate or subobtuse. Capsule only slightly longer than calyx. Leaves narrow-ovoid ....21. **S. merzbacheri** Ju. Kozhevn.

+   Sepals lanceolate, with narrow scarious margin. Leaves lanceolate or linear-lanceolate .... 32.

32.   Capsule equal to calyx or somewhat longer. Plant 20-40 cm tall. Petals 2/3 shorter than sepals ....29. **S. viridescens** (Maxim.) Ju. Kozhevn.

+   Capsule twice longer than calyx. Plant usually not taller than 15 cm. Petals longer than sepals ....33.

33.   Calyx 5-6 mm long. Leaves often falcate. Pedicels on fruit usually laterally deflexed. Peduncles somewhat thick, much shorter than leaf-covered portion of plant .... 9. **S. dahurica** Willd.

+   Calyx 3-4.5 mm long. Leaves straight. Pedicels on fruit not deflexed. Peduncles long, slender, only slightly shorter than or equal to leaf-covered portion of plant .... 23. **S. peduncularis** Bge.

34.   Petals 1/2-2/3 shorter than sepals, 5-8 mm long.. .... 5. **S. brachypetala** Bge.

+   Petals only slightly shorter than, equal to or somewhat longer than sepals, 3-6 mm long .... 35.

35.   Leaves 1.5-2.5 cm long. Sepals with 3 convex green nerves.. .... 15. **S. graminea** L.

+   Leaves 0.5-1.5 cm long. Nerves on sepals indistinct, not distinguished by colour .... 36.

36.   Petals somewhat shorter than calyx, 2.5-3 mm long. Cilia mainly at internodes.................... 32. **S. williamsiana** (Will.) Ju. Kozhevn.

+   Petals equal to calyx, 5-5.5 mm long. Internodes glabrous. Cilia mainly at leaf base .... 25. **S. pusilla** Schmid.

1. **S. alexeenkoana** Schischk. in Fl. URSS, 6 (1936) 882; Fl. Tadzh. 3 (1968) 480; Opred. rast. Sr. Azii [Key to Plants of Mid. Asia] 2 (1971) 232.

—*Mesostemma alexeenkoana* (Schischk.) Iconnik. (Ikonnik.). in Novit. syst. pl. vasc. 13 (1976) 114.

Described from East. Pamir (Tagdumbash-Pamir). Type in St.-Petersburg (LE). Plate I, fig. 5.

On rocky slopes and talus at about 4000 m alt. or higher.

IIIC. **Pamir** (In angustis Pistan, jugi Sary-bas, in lapidosis, July 15, 1901, Alexeenko—typus).

General distribution: Pamir.

2. **S. alsinoides** Boiss. et Buhse in Nouv. Mem. Soc. Natur. Moscou (1860) 41; Boiss. Fl. Or. 1 (1867) 705; Schischk. in Fl. SSSR, 6 (1936) 423; Fl. Tadzh. 3 (1968) 474. —*Tyttostemma alsinoides* (Boiss. et Buhse) Nevski in Acta Inst. bot. Ac. Sci. URSS, ser. 1, 4 (1937) 305; Fl. Turkm. 3 (1946) 22; Fl. Uzb. 2 (1953) 352; Opred. rast. Sr. Azii [Key to Plants of Mid. Asia] 2 (1971) 232; Kamelin et al in Byull. Mosk. obshch. isp. prir., otd. biol. 90, 5 (1985) 112. —**Ic.:** Fl. Turkm. 3, fig. 4.

Described from Iran. Type in Geneva (G).

On mountain slopes.

IIA. **Junggar:** *Jung. Gobi* (south. extremity of Khuvchiin-nuru mountain, No. 825, Aug. 1; nor. extremity of Maikhan-Ulan mountain, 28 km southwest of Bugat somon, No. 796, Aug. 1, 1984—Kamelin et al).

General distribution: Fore Asia, Mid. Asia.

3. **S. amblyosepala** Schrenk in Fisch. et Mey. Enum. pl. nov. 2 (1842) 54; Kryl. Fl. Zap. Sib. 5 (1931) 992; Trautv. in Bull. Soc. natur. Moscou, 32, 1 (1860) 159; Schischk. in Fl. SSSR, 6 (1936) 398; Opred. rast. Sr. Azii [Key to Plants of Mid. Asia] 2 (1971) 229; Grub. Konsp. fl. MNR [Conspectus of Flora of Mongolian People's Republic] (1955) 125; id. Opred. rast. Mong. [Key to Plants of Mongolia] (1982) 100. —*S. dichotoma* ε *rigida* Bge. Suppl. Fl. alt. (1836) 34; Fenzl in Ledeb. Fl. Ross. 1 (1842) 380. —*S. rigida* α *typica* Rgl. in Acta Horti Petrop. 5 (1877) 36.

Described from East. Kazakhstan (Jung. Ala Tau). Type in St.-Petersburg (LE). Plate II, fig. 5. Map 2.

In arid regions along river banks, on sand, rocky slopes, rocks.

IA. **Mongolia:** *Mong. Alt.* (west. extremity of Tsetseg-nur basin, Temetiin-Khukh-ula, south-west, slope on road from Tsetseg somon to Must somon, 2150-2200 m, June 26; Buyantu river basin, west. spur of Bugu-Ula on road from ford through Dzhangyz-Agach to Tugmentu-daba 5 km north of crossing, 2260 m, July2; interfluvine region of Kobdo river and Saksai-gol on road to Khargantu-gol, west. mountain trail to of Temir-Kherbein-nuru to south of Tsagan-Khada-ula, 1900 m, July 7, 1971—Grub., Ulzij. et al; Nyutsugun-gol, tributary of Uenchi-gol, 2100 m alt., along river bank, June 26, 1973—Golubkova, Tsogt; Bulgan-gola basin, Ded-Nariin-sala 1 km from estuary, 1550-1600 m alt., rocky slope (granites), Aug. 17, 1979—Grub., Dariima et al). *Depr. Lakes* (granite bald peak 3 km south of Ulangom on road to Khobdo, gorge, rocky slopes, July 25, 1945—Yun.), *Gobi Alt.* (Ikhe-Bogdo town, foothill and lower belt of mountains,

Aug. 18, 1926—Tug.; Noyan-ula mountain range, among rocks, July 19, 1973 —Golubkova, Tsogt; Bayan-Tukhum lake basin, south-east, extremity, flat, fixed sands, Sept. 10, 1979—Grub., Dariima et al.), *West. Gobi* (Atas-Bogdo town, along east. mountain trail and mountain slopes, Aug. 12-13, 1943—Tsebigmid), *Alash. Gobi* (Khuren-Khana mountain range, along Khailyasyn-khundii gorge near Elstiin-Dzadgai spring, alt. about 1600 m, rocks exposed north in fissures, Sept. 6, 1979—Grub., Dariima et al), *Khesi* (Chzhan'e 15 km north of Yunchan town, rocky slopes of Baidashan' mountain, June 28, 1958—Petr.).

IIA. **Junggar:** *Tarb.* (Dzhair mountain range, gorge north of Dzhair crossing and 4-5 km south of Yamata picket on road to Chuguchak, along rock cleavages, Aug. 4, 1957—Yun. et al), *Tien Shan* (Mulei area, Dashitou village, on slope, No. 2116, Sept. 24, 1957—Kuan), *Jung. Gobi* (Ikhe-Alag-ula mountains on Khairkhan somon—Altay somon road (Bayan-Obo) 18 km from gorge beginning, Khatsovchiin-bulak spring, canyon rocks exposed south, Aug. 20, 1979—Grub., Dariima et al).

General distribution: Fore Balkh., Jung.-Tarb.; West. Sib. (Altay).

4. **S. arenaria** Maxim. Fl. Tangut. 1 (1889) 91; Williams in Bull. Herb. Boiss. 2, ser. 10 (1907) 834; id. in J. Linn. Soc., London (Bot.) 38 (1907—1909) 397. —*S. decumbens* var. *acicularis* Edgew. et Hook. f. in Fl. Brit. India, 1 (1875) 235; Zhou Li-hou in Fl. Xizang. 1 (1983) 703. —Ic.: Maxim. Fl. Tangut. Plate 29, fig. 18-26; Fl. Xizang. 1, tab. 222, fig. 11-15.

Described from Tibet. Type in St.-Petersburg (LE). Plate 1, fig. 3.

On sandy shoals, alpine meadows, at 4300-5500 m alt.

IIIB. **Tibet:** *Weitzan* (Dzhagyn-gol river, on wet sand bank, 4000 m alt., July 2, 1900—Lad., typus !), *Chang Tang* (Zhitu, Zhou Li-hou, l.c.), *South.* ("Chzhada, Pulan', Chzhunba"—Zhou Li-hou, l.c.).

General distribution: Himalayas (east.).

Note. When treating *Stellaria* subgen. *Adenonema*, F. Williams (1907) had only a very vague idea about it and considered it as intermediate to *S. cherleriae* (Fisch.) Will. and *S. dicranoides* (Cham. et Schlecht.) Fenzl. Moreover, he regarded this species as closer to the latter, which view cannot be supported at all. He did not study the type specimen of *S. arenaria* but cited it in his article only on the basis of C. Maximowicz's diagnosis which he had copied almost completely and even added "Semina punctata" not mentioned in the original diagnosis as he did not have type specimens ("capsula atque semina nondum nota"—Maximowicz, 1889: 92). Nevertheless, the Herbarium of Komarov Botanical Institute (LE) has specimens (obviously received from F. Williams) of typical *S. arenaria* (Tibet Frontier Commission, coll. Younghusband, 1903), identified by F. Williams as *S. decumbens* var. *acicularis* Edgew. et Hook. f. This identification is essentially correct but, in his treatment, F. Williams placed this variety among synonyms of *S. cherleriae* var. *fasciculata* (Fisch.) Will. (with citation of Younghusband's specimens) but this completely differs from the basonym of this variety as well as from the author's concept of *S. decumbens* Edgew. s.l. which is far more different from *S. cherleriae* (Fisch.) Will. (*S. petraea* Bunge) than from *S. arenaria*. Obviously, his interpretation of *S. arenaria* Maxim. is not correct.

5. **S. brachypetala** Bge. in Ledeb. Fl. alt. 2 (1830) 161; id. in Mem. Ac. Sci.
St.-Petersb. 2 (1835) 548, cum var. *erecta;* Fenzl in Ledeb. Fl. Ross. 1 (1842)
390; Trautv. in Bull. Soc. natur. Moscou, 32, 1 (1860) 159; O. Fedtsch. in Tr.
Bot. muz. Akad. nauk, 7 (1910) 151; Hand.-Mazz. Symb. Sin. 7 (1929) 191;
Kryl. Fl. Zap. Sib. 5 (1931) 1000; Schischk. in Fl. SSSR, 6 (1936) 405; Grub.
Konsp. fl. MNR [Conspectus of Flora of People's Republic of Mongolia]
(1955) 125; Opred. rast. Sr. Azii [Key to Plants of Mid. Asia] 2 (1971) 230;
Grub. Opred. rast. Mong. [Key to Plants of Mongolia] (1982) 101; Ikonnik.
Opred. rast. Pamira [Key to Plants of Pamir] (1963) 106; Claves pl. Xinjiang.
2 (1983) 248. —*S. graminea* var. *brachypetala* Rgl. in Bull. Soc. natur. Moscou,
35, 1 (1862) 287; Maxim. Fl. Tangut. 1 (1889) 91; id. Enum. pl. Mong. 1
(1889) 102.

Described from Altay (Chuya river). Type in St.-Petersburg (LE).

In coastal and alpine meadows, rock screes and talus, in middle and
upper mountain belts.

IA. Mongolia: *Khobd.* (Dzusylan, in forest, July 13, 1879—Pot.; Khatu river, Boku-
Merina tributary, Aug. 6, 1909—Sap.; Achit-nur lake, marshy interfluvine region of
Bukhu-Murena and Khubusu-gola, 7-8 km off Bukhu-Muren somon, solonetzic sedge
meadow, July 15, 1971—Grub., Ulzij. et al (together with var. *magna*), **Mong. Alt.,**
(slope of Malyi Ulan-Daban crossing, between Dabasutai lake and Bodunchi river, July
19; Uruktu river bank, July 2; bank of Barsun-gol river, tributary of Kengurlen river, July
5, 1898—Klem.; Adzhi-Bogdo mountain range, Burgachin-dava crossing, between
Indertiin-gol and Dzuslangin-gol, rubble screes of alpine belt, Aug. 6; midportion of
Bus-Khairkhan mountain range, ravine with steep slopes, July 17, 1947—Yun.; Kobdo
river basin, Duro-nur lake, east. bank on road to Delyun, Buratiin-gol river, 2410 m alt.,
solonchak sedge meadow, June 30, 1971—Grub., Ulzij. et al; Nyutsugun-gol, Uenchin-
gol tributary, 2100 m alt., on river bank, June 26, 1973—Golubkova, Tsogt), **Val. Lakes**
(on sandy left bank of Tsagan-Turina river, not far from Uta river, June 14, 1894—
Klem.), **Gobi-Alt.** (Dundu-Saikhan mountains, around rocks in wet locations, July 9,
1909—Czet.; Ikhe-Bogdo mountain range, in a rocky bed of brook, Narin-Khurimt-ama
estuary, May 28, 1945—Yun.).

IIA. Junggar: *Tarb.* (Khobuk river valley, larch forest along brook, July 20, 1914—
Sap.; *Saur, Taz crossing, rocky alpine screes, June 21, 1914—Schischk., Genina), **Tien-
Shan** (on M. Yuldus, 1877—Przew.; Turkyul' lake, willow grove, June 16; Nan'shan-kou,
May 26, 1877—Pot.; Kumbel', nor. slope of Iren-Khabirga mountain range, May 31,
1879—A. Reg.; mountains near Santash crossing, in forests, June 10, 1893—Rob.; Advak
area, in forest, Aug. 29, 1895—Rob.; Sudliches Klukonik Tal, beim lager unter Tschon
Yailak Pass, June 15, 1908—Merzb.; Manas river basin, left bank, upper valley of Danu
river, on ascent to Se-daban crossing, high-mountain belt, semimarshy talus, July 21;
same locality, Danu-Daban crossing between Ulan-Usu valley and Danu-gol 1-2 km
west of crossing, semimarshy placers, July 23, 1957—Yun. et al).

IIIA. Qinghai: *Nanshan* ("along south, slope, 3400-3600 m alt., July 23, 1979,
Przew."—Maxim. l.c.).

IIIB. Tibet: *Weitzan* (Yangtze-Tszyan'a basin (Golubaya river), Chyuchen-Sum-do
area, on old Tangut pastures, 3600-3900 m alt., July 19, 1900—Lad.).

IIIC. Pamir (**in angustis Chargush, alt. 4200 m, Sept. 3, 1878, Paulsen;
Tagdumbasch, in Ilyk-su valley, loco Kaschka-su supra confl. Sararyk, July 17, 1901,
Alexeenko).

General distribution: Jung.-Tarb., Nor. Tien Shan, Cent. Tien Shan, East. Pam.;
West. Sib. (Altay), East. Sib. (Sayans), Nor. Mong. (Hang., Mong.-Daur.).

18

Note. In our territory, 2 varieties can justifiably be distinguished within this species: *S. brachypetala* var. *alatavica* (M. Pop.) Ju. Kozhevn. in Novit. syst. pl. vasc. 20 (1983) 106 and *S. brachypetala* var. *magna* Ju. Kozhevn. l.c. 106. The former, an alpine derivative of *S. brachypetala*, is characterized by smaller sizes of the plant as a whole and its constituent parts: proximated short stems forming a tuft. There is no distinct boundary between type and this variety. Under alpine and low-mountain intermediate conditions, both varieties grow together. In the list, var. *alatavica* is asterisked (*).

The second variety is distinguished by strong straight stem (up to 25 cm tall) and pedicels deflexed on fruit. In the list, this variety is double-asterisked (**).

6. **S. bungeana** Fenzl in Ledeb. Fl. Ross. 1 (1842) 376; Kryl. Fl. Zap. Sib. 5 (1931) 988; Schischk. in Fl. SSSR, 6 (1936) 395; Grub. Konsp. fl. MNR [Conspectus of Flora of People's Republic of Mongolia] (1955) 125; id. Opred rast Mong. [Key to Plants of Mongolia] (1982) 99. —*S. nemorum* Bge. in Ledeb. Fl. alt. 2 (1830) 152, non L.; Forbes and Hemsl. in J. Linn. Soc. (London) Bot. 23 (1886) 68. —*S. nemorum* β *bungeana* Rgl. in Bull. Soc. natur. Moscou, 35, 1 (1862) 268, cum a) *latifolia* and b) *angustifolia*.

Described from Siberia. Type in St.-Petersburg (LE).

In shady forests, humid scrubs in ravines.

IA. Mongolia: *Khobd.* (along nor.-east, descent of Ulan-Daban, in coniferous forest, July 22, 1879—Pot.; in Kharkhiry river valley, on rock screes, July 10, 1879—Pot.), *Mong. Alt.* ("Ulan-daba"—Grub. l.c.).

General distribution: Europe, West. Sib., East. Sib., Far East, Nor. Mong. (Hang., Hent.), China (Nor.), Korean peninsula, Japan.

7. **S. cherleriae** (Fisch. ex Ser.) Will. in Bull. Herb. Boiss., 2 ser. 10 (1907) 830, cum var. *uniflora*; id. in J. Linn. Soc. (London) Bot. 38 (1907–1909) 397; Kryl. Fl. Zap. Sib. 5 (1931) 1003; Schischk. in Fl. SSSR, 6 (1936) 420; Grub. Konsp. fl. MNR [Conspectus of Flora of People's Republic of Mongolia] (1955) 126; id. Opred. rast. Mong. [Key to Plants of Mongolia] (1982) 100; C.Y. Ma in Fl. Intramong. 2 (1978) 171. —*Arenaria* ? *cherleriae* Fisch. ex Ser. in DC. Prodr. 1 (1824) 409, cum α *uniflora*. —*S. petraea* Bge. in Ledeb. Fl. alt. 2 (1830) 160; Fenzl in Ledeb. Fl. Ross. 1 (1842) 395; Turcz. in Bull. Soc. natur. Moscou, 15 (1842) 608; Schischk. in Fl. SSSR, 6 (1936) 419; Grub. Konsp. fl. MNR [Conspectus of Flora of People's Republic of Mongolia] (1955) 127; id. Opred. rast. Mong. [Key to Plants of Mongolia] (1982) 100; Hanelt and Davazamc in Feddes Repert. 70, 1–3 (1965) 25; Opred. rast. Sr. Azii [Key to Plants of Mid. Asia] 2 (1971) 232. —*Adenonema petraeum* Bge. in Mem. Ac. Sci. St.-Petersb. Sav. Etrang. 2 (1835) 549. —*S. viridifolia* Pax ex Hoffm. in Feddes Repert. 12 (1922) 364; C.Y. Ma in Fl. Intramong. 2 (1978) 171. —**Ic.** Fl. Intramong. 2, tab. 89, fig. 1–3.

Described from East. Siberia (Dauria). Type in Geneva (G). Plate I, fig. 1. On sparsely grassy rocky slopes and peaks of mountains, on rocks.

IA. Mongolia: *Khobd.* (***on arid mountain at south. source of Kharkhira river, July 24, 1879—Pot.), *Mong. Alt.* (Tonkhil somon, Dzuilin-gol valley, crossing 2925 m north of Chindamani-ula, June 23, 1971—Grub., Ulzij. et al), *East. Mong.* (*on Suma-Khada mountain range, along rock screes, June 2, 1871—Przew., together with var. *intermedia*; Manchuria station, Beishan' mountain, June 23, 1951—S.H. Li et al; *Dariganga, Shiliin-Bogdo-ula, nor. slope, Aug. 11, 1970; Khalkha-gol somon, Numergin-gol, steppized slope, June 24, 1975—Zhurba), *Gobi-Alt.* (Baga-Bogdo, Narin-Khurimt creek valley, nor. slope, June 28, 1945; same locality, mountain steppe belt in contact zone with alpine carex groves, Sept. 11, 1943—Yun.).

IIIB. Tibet: *Weitzan* (*on rocks of Konchyun-chyu river, July 1; downstream along By-Dzhun river, July 6—1884, Przew.; **isthmus between Russkoe and Ekspeditsii lakes, in willow groves along interrupted mountain ravines, 4050 m, June 28, 1900—Lad.).

General distribution: Jung.-Tarb., Nor. Tien Shan; West. Sib., East. Sib., Nor. Mong.

Note. The following varieties can justifiably be cited within this species with synonymy.

var. *fasciculata* (Fisch.) Will., l.c., 833, excl. syn. *S. decumbens*; id. J. Linn. Soc. (London) Bot. 38 (1907–1909) 397. —*Arenaria* ? *cherleriae* γ *fasciculata* Bge. in Mem. Ac. Sci. St.-Petersb. 2 (1835) 549 —*S. petraea* Fenzl in Ledeb. Fl. Ross. 1 (1842) 394, cum α *vegeta* and β *tenuifolia*. —*S. petraea* var. *fasciculata* Bge. in Ledeb. Fl. alt. 2 (1830) 160; Maxim. Fl. Tangut. (1889) 92; id. Enum. pl. Mong. (1889) 102.

This variety is often represented by large plants in sharp contrast with other varieties. In the list, it has been asterisked (*).

var. *intermedia* (Turcz.) Ju. Kozhevn. in Novit. syst. pl. vasc. 21 (1983) 106. —*S. petraea* Bge. var. *intermedia* Turcz. in Bull. Soc. natur. Moscou, 15 (1842) 609.

This variety has linear-oblong leaves, pubescent stem and one or more flowers. In the list, it has been double-asterisked (**).

var. *alpina* (Bge.) Schischk. in Kryl. Fl. Zap. Sib. 5 (1931) 1003. —*Adenonema petraea* α *alpina* Bge. l.c., 549. —*Stellaria petraea* Bge. l.c., 160; Turcz. in Bull. Soc. natur. Moscou, 15 (1842) 609, α *alpina*. —*S. petraea* var. *imbricata* Fenzl in Ledeb. Fl. Ross. 2 (1842) 395; Maxim. Enum. pl. Mong. (1889) 102. —*S. cherleriae* var. *typica* Will., l.c., 831.

var. *uniflora* distinguished by Fischer, not var. *typica* Will. or var. *alpina* (Bge.) Schischk. is the type variety of this species. The latter 2 are synonyms of *S. petraea* distinguished by Bunge as emerging from De Candolle's "Prodromus" in which it was mentioned that Fischer distinguished 2 varieties of *Arenaria* ? *cherleriae* in the following order: var. *uniflora* and var. *fasciculata*. Consequently, the former is the type species.

Later, Bunge (1835, 549) equated *Arenaria* ? *cherleriae* α *uniflora* recognized by Fischer with his *Adenoma petraeum* β *cherleriae* which has extremely compactly pubescent stems and peduncles. However, the combination *S. cherleriae* var. *uniflora* (Fisch.) Will. in Bull. Herb. Boiss. 2, 7 (1907) 830; id, in J. Linn. Soc. (London) Bot. 38 (1907–1909) 398, although correct nomenclaturally, does not correspond to the available plant specimens of Fischer (LE). F. Williams' article also reveals that he did not interpret correctly the relations between races within *S. cherleriae* since he included *S. decumbens* var. *pulvinata* Edgew. and Hook. f. in the synonyms of this variety. The latter is not at all close to *S. cherleriae* var. *uniflora* and differs at once in the yellow-green unthickened margins of keel-shaped leaves as well as other characteristics that bring it close to *S. arenaria* Maxim. The inexact interpretation of the above taxa by F. Williams does not permit us to include the locations cited by him in our list. Williams' viewpoint was accepted by Pampanini (1930, see *S. maximowiczii*).

S. cherleriae was recognized by B.K. Schischkin (1936) as a species distinctly different from *S. petraea*. His conviction that these species are well distinguished was partly based on the material from Baikal regions, including mature steppe plants. However, the significant habitat differences between steppe and high-mountain plants do not yield any delimiting characteristics of diagnostic importance. Local populations reveal gradual morphological variations of general plant form associated with environmental changes.

The question of adequate isolation of Baikal and North Mongolian plant species remains debatable. When this is decisively resolved, epithet *cherleriae* will not be applicable to the new taxon.

8. **S. crassifolia** Ehrh. Hannov. Mag. 8 (1784) 116; Ser. in DC. Prodr. 1 (1824) 398; Bge. in Ledeb. Fl. alt. 2 (1830) 156; Fenzl in Ledeb. Fl. Ross. 1 (1842) 383; Turcz. in Bull. Soc. natur. Moscou, 15 (1842) 606; Maxim. Enum. pl. Mong. 1 (1889) 101; Kryl. Fl. Zap. Sib. 5 (1931) 994; Schischk. in Fl. SSSR, 6 (1936) 402; Grub. Konsp. fl. MNR [Conspectus of Flora of Mongolian People's Republic] (1955) 126; Opred. rast. Sr. Azii [Key to Plants of Mid. Asia] 2 (1971) 230; Grub. Opred. rast. Mong. [Key to Plants of Mongolia] (1982) 100; C.Y. Ma in Fl. Intramong. 2 (1978) 173; Claves pl. Xinjiang. 2 (1983) 248, incl. var. *linearis* Fenzl. —**Ic.:** Fl. Intramong. 2, tab. 92, fig. 1-3.

Described from Europe. Type in Berlin (?).

In moist sites on banks of rivers and lakes, swamps, moist meadows.

IA. Mongolia: *Mong. Alt.* (Khara-Adzarga mountain range, Sakhir-sala river valley, on moss cover in river water, Aug. 22; same site, valley of Shutyn-gol river, deep rocky gorge, on pebble bed near water, Aug. 28, 1930—Pob.; sources of Ubchugiin-gol, hummocky solonchak meadow with swampy hollows at spring mouths, Sept. 9, 1948—Grub.), *Depr. Lakes* (around Ubsa lake, in Kharkhira river valley, in quiet branch of river, Aug. 20; along swamp fringe around Dzeren-nor salt lake, Aug. 5, 1879; near Kirgiz-nor lake, Aug. 5—1879, Pot.; Kharkhira river valley, birch grove, Sept. 1, 1931—Bar. and Shukhardin; in Kaada river gorge entering Ubsa lake, wet pebbly banks, July 4, 1892—Kryl.), *Val. Lakes* (nor. bank of Bon-Tsagan lake west of Baidarik estuary, rushes-sedge solonchak meadow, June 16, 1971—Grub., Ulzij. et al), *East. Mong.* (on mountain slopes south-east of Ulankhanga, 1899—Pal.).

IIA. **Junggar:** *Cis-Alt.* (Qinhe-Chzhunkhaitszy, on water, No. 1390, 1374, Aug. 6, 1956—Ching).

General distribution: Arct., panboreal.

9. **S. dahurica** Willd. ex Schlecht. Mag. Ges. Naturf. Freunde, 7 (1816) 195; Ser. in DC. Prodr. 1 (1824) 399; Fenzl in Ledeb. Fl. Ross. 1 (1842) 388; Turcz. in Bull. Soc. natur. Moscou, 15 (1842) 600; Schischk. in Fl. SSSR, 6 (1936) 415; Grub. Konsp. fl. MNR [Conspectus of Flora of Mongolian People's Republic] (1955) 126. —*S. glauca* α *dahurica* Rgl. in Bull. Soc. natur. Moscou, 35 (1862) 293. —*S. graminea* δ *dahurica* Glehn ex Maxim. Enum. pl. Mong. (1889) 101; Kryl. Fl. Zap. Sib. [Flora of West. Siberia] 5 (1931) 997. —**Ic.:** Fl. Zabaik. 4, fig. 159.

Described from East. Siberia (Dauria). Type in Berlin (?).

In forests, along river banks, humus soils on mountain slopes.

IA. **Mongolia:** *Mong. Alt.* (Taishiri-ula mountain range, nor. slope, larch forest, July 12, 1945—Yun.).

IIA. **Junggar:** *Tien Shan* (prope Norin-Chorol, 1834, collector not known; Kumbel', May 30, 1879—A. Reg.; from Bortu to timber works in Khomote, in spruce forest, No. 7037, Aug. 4, 1958—Lee and Chu), *Jung. Gobi* (Ikhe-Khavtag mountain range on border with People's Republic of China in Dzeg post region, July 14, 1984—Gub.).

General distribution: East. Sib., Nor. Mong. (Fore Hubs., Hent. Hang., Mong.-Daur..).

10. **S. decumbens** Edgew. in Trans. Linn. Soc. (London), 20 (1851) 35; id. in Hook. f. Fl. Brit. India, 1 (1875) 234, var. *edgeworthii*; Zhou Li-hou, Fl. Xizang. 1 (1983) 702.

Described from Himalayas. Type in London (K).

On rocks, alpine meadows and coastal pebble beds, 4300–5600 m alt.

IIIB. **Tibet:** *South.* ("Tszyanda, Chzhunba"—Zhou Li-hou, l.c.).

General distribution: endemic.

11. **S. depressa** Schmid in Feddes Repert. 31 (1933) 41. —*S. decumbens* Edgew. in Hook. f. Fl. Brit. India, 1 (1875) 234, p.p. —*S. roborovskii* Ju. Kozhevn. in Novit. syst. pl. vasc. 20 (1983) 106.*

Described from West. Tibet. Type in Zurich (Z). Plate II, fig. 1; Plate IV, fig. 5.

On coastal sand and pebble beds, alpine meadows.

IB. **Kashgar:** *Nor.* ("Uch-turfan, June 18, 1908—Divn.).

IIA. **Junggar:** *Tien Shan* (6-7 km south of Danyu, on east. slope, No. 452, 486, July 22; between Danyu and Daban, No. 548, July 23, 1957—Kuan).

IIIA. **Qinghai:** *Nanshan* (nor. slope of Humboldt mountain range, alpine meadow, June 7, 1908—Rob., typus; same site, Kuku-usu river, between rocks on bank, June 6, 1908—Rob.).

IIIB. **Tibet:** *South.* (Aksai-Chin, 5000 m, Sept. 5; Tschu-sangpo, am Lanak-La, 5450 m, Aug. 13, 1937—Bosshard).

Stellarra roborowskii Ju. Kozhevn. was not described at all.—Translation editor

IIIC **Pamir** (on ascent to Billuli pass from Gumbus, in rocky valley of brook, June 12, 1909—Divn.).

General distribution: Himalayas (west.).

Note. Within the species, variety *S. depressa* var. *arenicola* Yu. Kozhevn. comb. nov. —*S. roborowskii* var. *arenicola* Yu. Kozhevn. in Novit. syst. pl. 20 (1983) 105—has been distinguished and asterisked (*).

We described *S. roborowskii* Ju. Kozhevn. in 1978 before we were aware of the material of V. Bosshard's expedition. In the material kindly supplied to us by Kew, there was a sheet from the herbarium of Hooker the son. These plants collected in Tibet at 4500 m alt. (Pass. n. Le, July 20, 1848, Coll. I.I.) were identified by R.R. Stewart in 1965 as "*S. palustris* Retz. running into *S. webbiana* Wall. ?". We identified these plants as *S. roborowskii*. In 1980, we received for information the plants collected by Bosshard and identified by E. Schmidt and preserved in Botanischer Garten und Institut fur Systematische Botanik der Universitat, Zurich. The identity of *S. depressa* and *S. roborowskii* has been established.

S. dianthifolia Will. in J. Linn. Soc. London (Bot.) 38 (1919) 396; Majumdar in Bull. Bot. Surv. India, 15, 1 (1973) 43; Zhou Li-hou in Fl. Xizang. 1 (1983) 698.

Described from Himalayas. Type in London (K). Isotype in Calcutta (Cal.).

General distribution: Himalayas (east.).

Note. Though the original description stated that the species was described from Tibet (Leinyaotszy), this territory falls in the Himalayas. However, in view of the proximity of Tibet, the occurrence of this species there is highly likely. This species is possibly of hybrid origin. Type specimens No. 366 (3 numbers) are mounted on the same sheet with specimens of hybrids *S. brachypetala* Bunge × ? *S. graminea* L.

For some unknown reason, F. Williams (1907–1909) did not mention in the original description that *S. dianthifolia* bears glandular pubescence.

12. **S. dichotoma** L. Sp. pl. (1753) 421; Ser. in DC. Prodr. 1 (1824) 397; Bge. in Mem. Ac. Sci. St.-Petersb. 2 (1835) 547, cum var. *cordifolia*; Fenzl in Ledeb. Fl. Ross. 1 (1842) 379, 380, cum var. *heterophylla*, var. *linearis*, var. *rigida*; Turcz. in Bull. Soc. natur. Moscou, 15 (1842) 600; Trautv. in Acta Horti Petrop. 1 (1871) 7; Forbes and Hemsl. in J. Linn. Soc. London (Bot.) 23 (1886) 67; Maxim. Enum. pl. Mong. 1 (1889) 99; Kryl. Fl. Zap. Sib. 5 (1931) 991; Schischk. in Fl. SSSR, 6 (1936) 397; Hao in Bot. Jahrb. 68 (1938) 594; Grub. Konsp. fl. MNR [Conspectus of Flora of Mongolian People's Republic] (1955) 126: id. Opred. rast. Mong. [Key to Plants of Mongolia] (1982) 100; Hanelt und Davazamc in Feddes Repert, 70, 1-3 (1965) 25; C.Y. Ma in Fl. Intramong. 2 (1978) 169; Fl. Desert. Sin. 1 (1985) 451. —*S. pallasiana* Ser. in

DC. Prodr. 1 (1824) 399. —*S. stephaniana* Willd. ex Schlecht. in Mag. Ges. Naturf. Freunde, 7 (1816) 194. —*S. dichotoma* var. *stephaniana* (Willd.) Rgl. in Bull. Soc. natur. Moscou, 35, 1 (1862) 273; Maxim. Fl. Tangut. 1 (1889) 100; Kryl. Fl. Zap. Sib. 5 (1931) 992. —*S. bistylata* W. Zh. Di et Y. Ren in Acta bot. bor.-occid. Sin. 5, 3 (1985) 231. —Ic.: Fl. SSSR, 6, Plate 9, fig. 4; Fl. desert. Sin. 1, tab. 165, fig. 1-5.

Described from Siberia. Type in London (Linn.). Plate II, fig. 3.

On rocky slopes, coastal pebble beds and shoals, flat and hummocky sands.

IA. **Mongolia:** *Khobd.* (around Ubsa lake, on sand bank of Bekon-bere river, June 17; in Uryuk-nora lake basin on Byrgusutai river, on pebbles, June 21; at Bairirmen-daban crossing, June 20, 1879—Pot.), *Mong. Alt.* (in Dzusylyn gorge, Adzhi-Bogdo peak, on rubble, June 29; on Urten-gol river emerging from Adzhi-Bogdo, on granite rocks, July 1; on rocks in Tsitsirin-gol river gorge, July 10, 1877—Pot.; nor. slope of Chinese Altay, pass [Tiektygish-Terekty], July 7, 1903—Gr.-Grzh.), *Cent. Khalkha* (around Ikhe-tukhum-nor lake, Ulan-Delger mountain, June; ravine between Bugen and Modto mountains, June 1926; rocks near Choiren, July 1, 1926—Pavlov; from Erga to Choiren, Sept. 2; around Baishintin-sume, Urgo mountain on way to Guntu khuduk, Aug. 18; from Choiren to Naran area on Sair-Usinsk road, Santa mountain, Sept. 9; Alkha-Khoshuni Gobi, Nabtsan mountain foothill, Aug. 14—1927, Zam.; on way from Ulan-Bator to Del'gir-Khangai mountains, on Khairkhan rocks, Aug. 28, 1931—Ik.-Gal.; old road to Dalan-Dzadagad on ascent to nor. slope of Gangyn-dava pass, Surtugchingin-nur area, along banks of small gorge, July 10, 1948—Grub.), *East. Mong.* (Kulun-buir-norsk plain, Kuitune-khara-saba, June 6, 1899—Pot. and Sold.; Bain-Tsakhto, pass, 1899—Pal.; Manchuria station, Beishan' mountain, June 23, 1951—S.H. Li et al; 70 km south-east of Choibalsan town, undulating plain, June 14, 1954; 30 km north of Ar-Zhargalant somon, June 28, 1956; 78 km nor.-east of Erdene-Tsagan, on basalt outcrops, July 1, 1971—Dashnyam), *Depr. Lakes* (around Khara-nor lake on nor. slope of Unyugut mountain, Aug. 14, 1879—Pot.; rubbly alluvium of Dundu-Tsinkir river, Aug. 15; Kharkhira river valley 6 km south of Ulangom, Aug. 6, 1930—Bar.), *Val. Lakes* (Kholt area in Hangay foothills, May 30, 1926—Gusev), *Gobi-Alt.* (Ondai Sair, Outer Mongolia, sand wash at 1680 m; Baga-Bogdo, on steep walls at 1800 m—1925, Chaney; Dundu Saikhan mountains, Aug. 17, 1931—Ik.-Gal.; south-west, slope of Ikhe-Bogdo mountain range, near estuary of Narin-Khurimt gorge, 2440-2500 m alt., July 30, 1948—Grub.), *East. Gobi* (on road to Khara-Tolit area from Mandashig-Gobi area, on sandy soil, 1909—Czet.; Dzamyn-ude, on weathered granite rocks, Aug. 26, 1931—Pob.), *Alash. Gobi* (on Alashan mountain range, alpine meadows, July 4, 1873—Przew.).

IIA. **Junggar:** *Tien Shan* (on rocks around Nanshan-kou, June 7, 1877—Pot.).

IIIA. **Qinghai:** *Nanshan* (on Kukunor mountain range, June 7, 1880—Przew.), *Amdo* ("Kokonor, auf dem plateau Dahopa, 4000 m, No. 1054, Aug. 28, 1930"—Hao, l.c.).

IIIC. **Pamir:** "Tagdumbash-Pamir, in Pistan gorge, Sary-Kol mountain range, on rocks, July 15, 1901, Alekseenko"—O. Fedtsch. 1910 (sub *S. rigida* Bge.).

General distribution: West. Sib. (Altay), East. Sib. (Sayan), Nor. Mong., China (Nor.-West.).

Note. Even A. Bunge (1835, 547) pointed to the inconsistency of characteristics of this species which makes it difficult to distinguish distinct varieties within it. Nevertheless, Bunge isolated 5 varieties, var. *cordifolia* representing the type.

In the Central Asian material available to us, only 1 variety of A. Bunge, namely var. *lanceolata*, is well-marked. As already pointed out, it is a hybrid produced by the hybridization of *S. dichotoma* and *S. gypsophylloides* Fenzl. This variety inherited from the former species a calyx with up to 4.5 m long lanceolate sepals and relatively large capsule with 2 (not 1) seeds; in some capsules, only 1 seed was fully formed while the second remained immature. This variety is closely related to the latter species in the absence of glandular pubescence in the terminal parts of plants. The absence of this pubescence primarily distinguishes *S. gypsophylloides* from *S. dichotoma* but, nevertheless, some stray stalked glandules can be detected even in entirely typical specimens of *S. gypsophylloides*, thus suggesting early hybridization, since sepals in this species may be sharp (as in *S. dichotoma*) and blunt, with broad membranous margin (as in typical *S. gypsophylloides*) within the same specimen. Since glandular pubescence is the most reliable characteristic for differentiating these species, the name of the hybrid should be derived from *S. gypsophylloides* (see below).

In our view, *S. bistylata* is a race of *S. dichotoma* since plants with 3 styles and 6-toothed capsules can be found, though rarely, among plants with 2 styles and 4-toothed capsules; in fact, *S. bistylata* is indistinguishable from *S. dichotoma* s. str.

13. **S. dilleniana** Moench, Enum. pl. Hass. (1777) 214; Schmid in Feddes Repert. 31 (1933) 41. —*S. fontana* M. Pop. ex Schischk. in Fl. SSSR, 6 (1936) 881; Ikonnik. Opred. rast. Pamira [Key to Plants of Pamir] (1963) 106; Fl. Tadzh. 3 (1968) 476; Opred. rast. Sr. Azii [Key to Plants of Mid. Asia] 2 (1971) 231; Podlech, Angers in Mitt. Bot. Staatssaml. Munchen, 13 (1977) 417. —Ic.: Fl. Tadzh. 3, Plate 77, fig. 1.

Described from Europe. Type ?

On banks of rivulets and brooks, on alpine grasslands in upper mountain belt.

IB. **Kashgar:** No. 114—V. Roborowsky's collection (1889) without date or site of collection.

IIA. **Junggar:** *Tien Shan* (Ala Tau mountains, Andzhelyav gorge on Khorgos river (25 versts—1 verst = 1.067 km—beyond Bashkunchi village, June 1907—Divn.).

IIIB. **Tibet:** *South.* (Keptung-La, July 29, 1927, Bosshard—Schmid, l.c.).

General distribution: Tien Shan, East. Pam.; Europe, East. Sib., Himalayas.

Note. A further study of this species may reveal it only as an intraspecific category of *S. palustris* Retz. In that event, *S. dilleniana* will be the type variety and the customary epithet *palustris* acquires an intraspecific rank.

14. **S. divnogorskajae** Ju. Kozhevn. in Novit. syst. pl. vasc. 20 (1983) 103. —*S. decumbens* Edgew. et Hook. f. in Hook. f. Fl. Brit. India, 1 (1875) 234, p.p.; O. Fedtsch. Tr. Bot. muz. Akad. nauk, 7 (1910) 151.

Described from Sinkiang (Kashgar). Type in St.-Petersburg (LE). Plate 11, fig. 2. Map 1.

IB. Kashgar: *Nor.* (Ucz-Turfan, Usch-Karagalik, June 18, 1908—Divn., typus).
IIIC. Pamir: Tagdumbash-Pamir, in angustis Pistan jugi Sarykol in lapidosis, July 15, 1901—Alexeenko).

Note. Var. *pilosa* Yu. Kozhevn. (l.c., 104) cited above for Tagdumbash-Pamir, has been distinguished within this species.

15. **S. graminea** L. Sp. pl. (1753) 422; Bge. in Ledeb. Fl. alt. 2 (1830) 159; Fenzl in Ledeb. Fl. Ross. 1 (1842) 391, cum var. *linearis* and var. *lanceolata*; Turcz. in Bull. Soc. natur. Moscou, 32, 15 (1842) 603; Trautv. in Bull. Soc. natur. Moscou, 1 (1860) 159, var. *ciliata*; Edgew. et Hook, f. Hook. f. in Fl. Brit. India, 1 (1875) 233; Hemsl. in J. Linn. Soc. (London) Bot. 23 (1886) 68; Diels, Fl. C. China (1901) 320; Hemsl. in J. Linn. Soc. (London) Bot. 35 (1902) 169; Diels in Futterer, Durch Asien (1903) 9; Will. in J. Linn. Soc. (London) Bot. 38 (1907–1909) 396; Kryl. Fl. Zap. Sib. 5 (1931) 404; Hao in Bot. Jahrb. 68 (1938) 594; Walker in Contribs. US Nat. Herb. 28, 4 (1941) 614; Grub. Konsp. fl. MNR [Conspectus of Flora of Mongolian People's Republic] (1955) 126; Opred. rast. Sr. Azii [Key to Plants of Mid. Asia] 2 (1971) 230; Grub. Opred. rast. Mong. [Key to Plants of Mongolia] (1982) 101; Hanelt and Davazamc in Feddes Repert. 70, 1–3 (1965) 25; Zhou Li-hou in Fl. Xizang. 1 (1983) 702. —*S. graminea* var. *laxmannii* (Fisch.) Trautv. in Acta Horti Petrop. 1 (1871) 7, non Fisch. —*S. gramineoides* J. Hazit, nom nud. in Claves pl. Xinjang. 2 (1983) 249.

Described from Europe. Type in London (Linn.).

Forest borders, among shrubs, meadows, rivers, shaded rocks, grassy mountain slopes up to 4800 m alt.

IA. Mongolia: East. Mong. (Khalkha-gol somon, quarry in Khamar-Daba, June 27, 1975—Zhurba; around Khailar town, swamps on hillocks, June 20, 1951—S.H. Li et al.), **Gobi-Alt.** (Dzun-Saikhan mountain, on rocks on wet terrace, Aug. 23, 1931—Ik.-Gal.; Ikhe-Bogdo-ula mountain range, south-east. slope, Narin-Khurimt gorge, on rocks, 2900 m alt., July 28, 1948—Grub.), **Alash. Gobi** (*Alashan mountain range, Yamato gorge, June 13, 1908—Czet.).
IB. Kashgar: South. (nor. slope of Takhtakhon mountains, on rocks, June 21, 1889—Rob.), **West.** (King-Tau mountain range, nor. slope 6-7 km south of Kosh-Kulak settlement, juniper-*Cobresia* spruce forest at 3400 m alt. (var. *montioides*); same site, steppe belt, in ravines on nor.-west, slope, in juniper beds (typical form), July 10, 1959—Yun. et al).
IIA. Junggar: Tarb. (Saur mountain range, south. slope, valley of Karagaitu river, Bain-Tsagan creek valley right bank, subalpine meadow, June 23, 1957—Yun. et al), **Jung. Alat.** (Syata-Ven'tsyuan', in forest, No. 1415, Aug. 13; Toli area, Koket-Denbas', nor. slope, in spruce forest, No. 1258, Aug. 6; Toli area, Barlyk mountains, on slope, No. 1021, Aug. 6, 1957—Kuan), **Tien Shan** (Mengute, Aug. 2; Bagaduslun, June 4, Usinchinkho, Apr. 29, 1879—A. Reg.; Urumchi, in lake basin, No. 0528, July 21, 1956—Ching; 36 km south-east of Nyutsyuan'tszy, No. 3, July 19; south of Nyutsyuan'tszy,

No. 553, July 16; nor. slope of Datszymyao, in Savan area, No. 440, July 22; around Barkul' lake, No. 2205, Sept. 27; 25 km south-east of Nyutsyuan'tszy town, on Nin'tszyakhe river bank, No. 57, July 17; in Savan area, Datszymyao, steppe, No. 1250, July 8; on Danukhe river, slope, No. 2190, July 22—1957, Kuan), *Jung. Gobi* (in Paotai state farm, No. 3548, Oct. 9, 1956—Ching).

IIIA. Qinghai: *Nanshan* (*near Kuku-nor lake, June 7, 1880; north of Tetung mountain range, July 4, 1872—Prezw.), *Amdo* (*Kokonor, auf dem Plateau Dahojba, 4000 m, No. 1054, Aug. 28, 1930"—Hao, l.c.).

IIIB. Tibet: *South.* ("Khamba Fort, 1903, Younghusband"—Williams, l.c.; Lhasa district, May-July 1947—Guthrie; "Tszyantsze, Tsyuishui"—Zhou Li-hou, l.c).

General distribution: panboreal.

Note. In our territory, 2 varieties are well distinguished within the species:.

var. *pilosula* Maxim. Fl. Tangut. 1 (1889) 91; id. in Acta Horti Petrop. 11, 1 (1889) 72; Walker in Contribs. US Nat. Herb. 28, 4 (1941) 614; Zhou Li-hou in Fl. Xizang, 1 (1983) 702. —**Ic.:** Fl. Xizang. 1, tab. 221, fig. 16-19.

Differs from type variety in stem with sparse pubescence of crispate hairs, cilia on entire margin of some leaves (apart from base) and petals shorter than sepals. Besides Qinghai, this variety is known in Himalayas (Dzjali, Nealamu) where it grows in forests and their edges at 3000–3500 m alt. (Zhou Li-hou, l.c.). It has also been reported in a bog in Mongolian Altay 10 km west of Khoton-nur, Aug. 17, 1979—Gubanov.

var. *montioides* Edgew. and Hook. f. Fl. Brit. India, 1 (1875) 233.

High-mountain variety with stunted, often creeping stems, with leaves shorter and less cuspidate than in type variety. Grows in Tibet at 4500–5200 m alt.

16. **S. gyangtsensis** Will. in J. Linn. Soc. (London) Bot. 38 (1907–1909) 396; Majumdar in Bull. Bot. Surv. India, 15, 1 (1973) 41.

Described from Tibet. Type in London (K). Isotype in St.-Petersburg (LE). Plate VI, fig. 4. Map 1.

On rocky sections in high mountains.

IIIB. Tibet: *South.* (Gyangtse, 1904, Walton-typus!; hill behind Gyangtse, alt. 4500 m, Aug. 13, alt. 4200 m, Aug. 16—1936, Gobshi; 2 miles south-east of Lhasa on low ground across Kye Chu, at 3600-3900 m, Sept. 1, 1936—Chapman).

General distribution: Himalayas (Sikkim).

Note. F. Williams (l.c.) pointed to similarity to *S. uda* in flower arrangement and leaf form. However, he did not mention that although *S. gyangtsensis* has 5-merous calyx, in some flowers, it appears 4-merous since 1 sepal appears as though pushed between calyx and petals. Thus, on the example of this species, a transition from 5- to 4-merous flower structure can be seen in genus *Stellaria*.

The observation of Williams that *S. gyangtsensis* has no pubescence is not correct as authentic material reveals extremely abundant pubescence on some internodes while some leaves bear cilia near base.

17. **S. gypsophylloides** Fenzl in Ledeb. Fl. Ross. 1 (1842) 380; Maxim. Enum. pl. Mong. 1 (1889) 100; Grub. Konsp. fl. MNR [Conspectus of Flora of People's Republic of Mongolia] (1955) 126; id. Opred. rast. Mong. [Key to Plants of Mongolia] (1982) 99; Hanelt und Davazamc, Feddes Repert. 70, 1-3 (1965) 25; Fl. desert. Sin. 1 (1985) 452. —*S. dichotoma* var. *lanceolata* (Bge.) Ma in Fl. Intramong. 2 (1978) 169. —Ic.: Grub. Opred. rast. Mong. [Key to Plants of Mongolia] Plate XI, fig. 192; Fl. desert. Sin. 1, tab. 165, fig. 6-7.

Described from Mongolia. Type lost. Lectotype in St.-Petersburg (LE). Plate II, fig. 4.

On sandy and clayey barren lands, arid rocky slopes and rocks, flanks and floors of gorges.

IA. Mongolia: *Mong. Alt.* (Khara-Adzarga mountain range, Khairkhan-Duru river vicinity, Aug. 26; same site, Shutyn-gol river valley, Aug. 28—1930, Pob.), *Cent. Khalkha* (clayey barren land, Aug. 19-31, 1873—Przew.), *East. Mong.* (*Mongolia chinensis, 1842—Gorski; *Dalai-nursk plain, near Khalun'-arshan, 1899—Pal.; *Shilin-khoto town, 1959—Ivan.), *Gobi-Alt.* (on nor. slope of Ubten-Kotel pass, Aug. 30; near Urdzhyum well, in same-named valley, Aug. 16; on nor. slope of Bain-Tsagan mountain range, July 23—1886, Pot.; on south, slope of Bayan-Tsagan mountains, Aug. 4; Bayan-Tukhum area, hummocky sand west of lake, Sept, 15—1931, Ik.-Gal.; nor. slope of Tszolen mountain range and its nor. trail covered with thin sand, July-Aug.—M. Simuk.; Bulgan-ula, south. of Barun-Tsokhe mountains rocky slopes, April 26, 1941; Nemegetu-nuru mountain range (west. tip), Khara-obo summit, Aug. 7, 1948; *Tostu-nuru mountain range, main summit of Sharga-morite, 2300—2565 m, Aug. 15—1948, Grub.; *Khurkhe-ula mountain range, rocky slopes of Altyn-Ama gorge, Sept. 7, 1950—Lavrenko et al; hummocky sand between Tszolin and Bayan-Tsagan towns, south of Bayan-Tukhum lake, Aug. 6, 1963—Petr.; 90 km south-west of Nomgon, near Sugyn-bulak spring, in gorge, July. 17; 32 km west-south-west of Gurvan-Tes settlement, Tostu-nuru mountain range, 2205m alt., under rocks, July 27; Khurkhe mountain range, 55 km south-east of Nomgon, on nor. slope, July 13—1972, Guricheva), *East. Gobi* (Ongon-Elisu sand, on sand mounds, Sept. 14; Kalgan road, not far from Sengei well, semidesert, Aug. 22; Dariganga, 40 km nor.-east of Argaleul mountain, Sept. 4; 17 km nor. of Dzamyin-Ude, on Khukh-Tologoi mountain slopes, Aug. 28—1931, Pob.; Sain-usu basin 20-25 km nor.-nor.-east of Sain-Shanda, Aug. 28, 1940; 60 km east of Ulan-Badarkhu somon on old caravan road, extensive sand mass, June 19, 1941—Yun.; 16 km nor.-east of Sain-Shanda on road to Baishintu, July 24, 1940—Shubin; Galbyn-Gobi vicinity, Nomogon-ula, rocky slopes of mountains, Aug. 1; Argalant mountains, Tsagan-obo cone-shaped hillock near Bayan-Munku-khida debris, 1000 m alt., Aug. 6—1970, Grub. et al; 60 km south-south-east of Khubsugul, Khutag-ula mountain, July 28, 1971—Isach.), *West. Gobi* (Atas-Bogdo-ula, upper mountain belt, Aug. 12; Tsagan-Bogdo mountain range, upper moiintain belt, Aug. 4—1943, Yun.; Tsagan-Bogdo mountain range, gorge 13 km on road to Ekhin-gol from Tsagan-bulak spring, gorge floor at 1850 m alt.; same site, under main peak at Sudzhiin-Bulak spring, at 2000—2380 m alt., in rock crevices in ravine, Aug. 29—1979, Grub. et al), *Alash. Gobi* (in Khalkhasov land, on clay with pebble, Aug. 31; south. Alashan mountains, Aug. 27—1873, Przew., Tszi-Tszi-kho area, in barren land, sand dunes, Sept. 23, 1901—Lad.; Tengeri sand, Shangyn-dalai area, July 8, 1909—Czet.; Inchuan', somewhat sand-covered, July 23, 1957; Bayan-Khoto, Tengeri sand, Dzhonamu, Aug. 15, 1958—Petr.), *Ordos* (Khonuikho river bank with shifting dunes, Aug. 6; Huang He, Aug. 6; Bayan-tokhum, on sand, Sept. 26; Taitukhai area, Aug. 29, 1884—Pot.; Ikochzhoumen, Ushinchi town, somewhat sand-covered Aug. 3, 1957—Petr.).

28

General distribution: endemic.

Note. *S. gypsophylloides* var. *lanceolata* (Bge.) Yu. Kozhevn. comb. nov. —
S. dichotoma lanceolata Bge. Fl. Alt. suppl. (1836) 34 (see note under *S. dichotoma* L.)—has been asterisked (*).

18. **S. irrigua** Bge. in Mem. Ac. Sci. St.-Petersb. 2 (1835) 548; Kryl. Fl. Zap. Sib. 5 (1931) 1002; Schischk. in Fl. SSSR, 6 (1936) 409; Ikonnik. Opred. rast. Pamira [Key to Plants of Pamir] (1963) 279; Fl. Tadzh. 3 (1968) 430. —*S. umbellata* Turcz. in Bull. Soc. natur. Moscou, 11 (1838) 5, nom. nud.; id. in Bull. Soc. natur. Moscou, 15 (1842) 607; Kar. et Kir. in Bull. Soc. natur. Moscou, 15, 1 (1842) 173; Fenzl in Ledeb. Fl. Ross. 1 (1842) 394; Maxim. in Fl. Tangut. 1 (1889) 91; Kryl. Fl. Zap. Sib. 5 (1931) 1002; Schischk. Fl. SSSR, 6 (1936) 409; Hao in Bot. Jahrb. 68 (1938) 594; Grub. Konsp. fl. MNR [Conspectus of Flora of Mongolian People's Republic] (1955) 127; Opred. rast. Sr. Azii [Key to Plants of Mid. Asia] 2 (1971) 229; Grub. Opred. rast. Mong. [Key to Plants of Mongolia] (1982) 100; Zhou Li-hou in Fl. Xizang. 1 (1983) 698; Claves Pl. Xinjiang. 2 (1983) 249. —*S. subumbellata* Edgew. in Hook. f. Fl. Brit. India, 1 (1875) 233; Hemsl. Fl. Tibet (1902) 169; Will. in J. Linn. Soc. (London) Bot. 38 (1907–1909) 396; Hand.-Mazz. Symb. Sin. 7 (1929) 192; Majumdar in Bull Bot. Surv. India, 15, 1 (1973) 40; Zhou Li-hou in Fl. Xizang. 1 (1983) 698. —Ic.: Fl. Xizang. 1, tab. 224, fig. 5–6.

Described from West. Siberia (Altay). Type in St.-Petersburg (LE). Map 2.

Around springs, along banks of rivers and lakes, on alpine meadows, 3000–5000 m alt.

IA. **Mongolia:** *Mong.-Alt.* ("Khara-Adzarga"—Grub. l.c.).

IB. **Kashgar:** *West.* (Sarykol' mountain range, westward of Kashgar, Bostan-Terek locality, July 11, 1929—Pop.).

IIA. **Junggar:** *Tien Shan* (Aktash in Dzhagastai mountains, Aug. 16, 1877; Arystyn, July 13; Kash river valley, south. slope, Aug. 17–18, 1879—A. Reg.; on Nilki-Dzinkho road via Yakou, on slope, No. 4083, Sept. 1, 1957—Kuan).

IIIA. **Qinghai:** *Nanshan* (Kuku-Usu area, around spring at 3300–3600 m alt., July 26, 1879—Przew.).

IIIB. **Tibet:** *South.* (Gooring Valley, at 5030 m, July, Aug., 1895—Littledale), *Weitzan* (mountain isthmus between Russkoe and Ekspeditsii lakes, 4200 m, June 28; Yantszy-tszyan river basin, Chokabu-vrun area, 4200 m, on gorge slopes, July 9, 1900—Lad.; "Amne Matchin, in der Nahe der Schneegrenze, No. 1155A, Sept. 3, 1930"—Hao, l.c.; "An'do"—Zhou Li-hou, l.c.).

General distribution: Arct. (Asian), West. Sib., East. Sib., Far East, Nor. Mong. (Fore Hubs., Hent., Hang., Mong.-Daur.), Himalayas (east., Kashmir), Nor. Amer.

Note. Twelve years after Bunge described *S. irrigua* Bunge on the basis of underdeveloped plant specimens, Turczaninow redescribed it as *S. umbellata* Turcz. The type of *S. irrigua* is preserved in Komarov Botanical Institute (LE) and not in Paris as mentioned by B.K. Schischkin in "Flora SSSR" (1936). Strangely, however, apart from the type herbarium sheet in this huge herbarium, there is one more sheet of S.S. Ikonnikov from Pamir, identified as *S. irrigua*. This specimen is no different from *S. umbellata* Turcz.; its type is preserved here and represented by several specimens from different regions of its vast distribution range.

Turczaninow was not aware of the specimens of *S. irrigua* when describing *S. umbellata* and, based on the description of Bunge, stated that the new species was very close to *S. irrigua*, differing from the latter only in the presence of petals. However, *S. irrigua* bears very small, as though vanishing, petals. It is known that some *Stellaria* species, especially with small petals, may have developed as well as obsolete petals. For this reason alone, the presence or absence of fine petals cannot serve as a characteristic for distinguishing the 2 species of *Stellaria*. Some plants, for example, from Czokabu-vrun area (V.F. Ladygin's collection, 1900) have flowers with and without petals. All the other differences in the type specimens of *S. irrigua* fall quite within the range of intraspecific variation of *S. umbellata*. It may also be added that there are no differences in geographical distribution and ecology between these 2 species.

Thus, both the names pertain to the same species. According to the rule of priority, Bunge's epithet, namely *S. irrigua*, should be retained.

S. subumbellata Edgew. (typus vidi) described from Himalayas and Tibet, according to Edgeworth, differs from the closely related *S. umbellata* Turcz. from Baikal in the smaller number of bracts and subtubercular seeds. It has also been pointed out that flowers of *S. subumbellata* are without petals but with 5 stamens.

C. Maximowicz (1889) too did not recognize Kashmir specimens of Aitchison as different from *S. umbellata* although they were identified as *S. subumbellata*. He rightly pointed out that the number of stamens in *S. umbellata* varied from 5 (when all of them are with anthers) to 10 (when some filaments are without anthers). Specimens of *S. subumbellata* collected by V.F. Ladygin and Littledale in Tibet, according to N.A. Turczaninow, are only small plants of *S. umbellata*.

Thus, the original description of Edgeworth (in Hooker f., 1875) as well as the evidence of Maximowicz and the specimens preserved in the Herbaria of Komarov Botanical Institute (LE) and Kew, lead to the conviction that *S. subumbellata* is not a distinct species but should be treated among synonyms of *S. irrigua* after *S. umbellata*.

S. lanata Hook. f. Fl. Brit. India, 1 (1875) 232; Will. in J. Linn. Soc. (London) Bot. 38 (1907–1909) 397; Zhou Li-hou, in Fl. Xizang. 1 (1983) 698.

Described from Himalayas (Sikkim). Type in London (K).

The type of habitat is not known.

General distribution: Himalayas (west.).

Note. Although this species has been cited by F. Williams (l.c.) in his article on Caryophyllaceae of Tibet, specimens from Tibet itself are not known so far and those cited by Williams and Zhou Li-hou pertain to the southern slope of Himalayas ("Redong-Oong, near Chumbi, 1878, Dungboo; Bang Chin, mountain west of Chumbi, 1882, King's Collector"—Williams, l.c.; "Lundze; Yadong"—Zhou Li-hou, l.c.; Becnylturi, July 17, 1882, King's Collector). The occurrence of this species in South. Tibet is, however, highly probable.

19. **S. maximowiczii** Yu. Kozhevn. in Novit. syst. pl. vasc. 20 (1983) 104. —*S. decumbens* Edgew. var. *pulvinata* Edgew. et Hook. f. in Hook. f. Fl. Brit. India, 1 (1875) 235; Maxim. Fl. Tangut. 1 (1889) 92; Hemsl. Fl. Tibet (1902) 169; Stewart in Bull. Torrey Bot. Club (1916) 631; Hedin, S. Tibet (1922) 84, Zhou Li-hou, Fl. Xizang, 1 (1983) 702. —*S. cherleriae* auct. non Fisch.: Pamp. Fl. Caracor. (1930) 104, non Fisch. —**Ic.:** Fl. Xizang. 1, tab. 223, fig. 6–10. Described from Himalayas. Type in London (K). Plate 1, fig. 4. Map 1. On rocky slopes of mountains, alpine meadows, 4500–5600 m.

IIIB. Tibet: *Chang Tang* (left bank of Yantszy-tszyan river, rocky placers of mountains, May 20, 1884—Przew.; "Banyge"—Zhou Li-hou, l.c.), **Weitzan** (on Konchyun-chyu river, June 30, 1884—Przew.; N.E. Tibet, between camp XXVII and camp XXVIII, 4849 m, Sept. 14, 1896"—Hedin, l.c.; "An'do"—Zhou Li-hou, l.c.), **South.** ("Gooring valley, 90°25', 30°12', about 1600 ft., Littledale"—Hemsley, l.c.; "Dinzhu"—Zhou Li-hou, l.c).
General distribution: Himalayas (west., Kashmir).

Note. While elevating the variety of Edgeworth and Hooker to the rank of species, its epithet should be changed as it was used by V.I. Grubov (1972) in the original description of another species (see below). *S. maximowiczii* is closest to *S. arenaria* Maxim., differing from the latter primarily in the smaller sizes of all parts of the plant and pulvinoid growth form.

20. **S. media** (L.) Cyr. Ess. pl. char. comm. (1784) 36; Vill. Hist. Pl. Dauph. 3 (1789) 615; Bge. in Ledeb. Fl. alt. 2 (1830) 153; Fenzl in Ledeb. Fl. Ross. 1 (1842) 377; Turcz. in Bull. Soc. natur. Moscou, 15 (1842) 599; Henders. and Hume, Lahore to Jarkand (1873) 312; Edgew. et Hook. f. in Hook. f. Fl. Brit. India, 1 (1875) 230; Forbes and Hemsl. in J. Linn. Soc. (London) Bot. 23 (1866) 68; Maxim. in Acta Horti Petrop. 11, 1 (1889) 71; Hand.-Mazz. Symb. Sin. 7 (1929) 188, pro var. *trientaloides* Hand.-Mazz.; Kryl. Fl. Zap. Sib. 45 (1931) 990; Schischk. in Fl. SSSR, 6 (1936) 395; Grub. Konsp. fl. MNR [Conspectus of Flora of Mongolian People's Republic] (1955) 127; Opred. rast. Sr. Azii [Key to Plants of Mid. Asia] 2 (1971) 229; Podlech, Angers in Mitt. Bot. Staatssaml. München, 13 (1977) 418; Ma Yu-ch. in Fl. lntramong. 2 (1978) 167; Zhou Li-hou in Fl. Xizang. 1 (1983) 692; Claves. pl. Xinjiang. 2 (1983) 248; Fl. desert. Sin. 1 (1985) 453. —*Alsine media* L. Sp. pl. (1753) 272. —**Ic.:** Fl. SSSR, 6, Plate 20, fig. 2; Fl. Tadzh. 3, Plate 77, fig. 6–7; Fl. Intramong. 2, tab. 89, fig. 4–8. Described from Europe. Type in London (Linn.).

On meadow slopes, up to 3900 m.

IIA. Junggar: *Tien Shan* (in Savan area, Shichan, Oct. 1, 1956—Ching).
IIIB. Tibet: *South.* ("Lhasa"—Zhou Li-hou, l.c.).

21. **S. merzbacheri** Ju. Kozhevn. in Novit. syst. pl. vasc. 20 (1983) 105. Described from Sinkiang (Tien Shan). Type in St.-Petersburg (LE). Plate III, fig. 1. Map 1.

IIA. Junggar: *Tien Shan* (opp. Urumtschi, lager am sudrande des Bogdo-ola, Aug. 26–29, 1908—Merzb., typus !).
General distribution: endemic.

S. monosperma Buch.-Ham. in D. Don, Prodr. Fl. Nepal (1823) 215; Majumdar in Bull. Bot. Surv. India, 15, 1 (1973) 43; Catal. Nepal. vasc. pl. (1976) 47. —*S. crispata* Wall. in Edgew. et Hook. f. in Hook. f. Fl. Brit. India, 1 (1874) 229; Zhou Li-hou in Fl. Xizang. 1 (1983) 691.

Described from Himalayas. Type in London (K). Isotype in Calcutta (Kolkata) (Cal).

On mountain meadows, in undergrowth.

Note. The occurrence of this species is probable in South Tibet. Only the following variety has been cited for adjoining Himalayan territory: var. *paniculata* (Edgew.) Majumdar in Journ. Ind. Bot. Soc. 44, 1 (1965) 141. —*S. paniculata* Edgew. in Trans. Linn. Soc. 20 (1867) 35; id. in Hook. f. Fl. Brit. India, 1 (1874) 229. This variety differs from the type in the length of calyx— 3–3.5 mm (not 4.5–5 mm) (Majumdar, l.c.).

22. **S. palustris** Retz. Fl. Scand. Prodr. ed. 2 (1795) 106; Forbes and Hemsl. in J. Linn. Soc. (London) Bot. 23 (1886) 68; Kryl. Fl. Zap. Sib. 5 (1931); Schischk. in Fl. SSSR, 6 (1936) 406; Grub. Konsp. fl. MNR [Conspectus of Flora of Mongolian People's Republic] (1955) 127; Opred. rast. Sr. Azii [Key to Plants of Mid. Asia] 2 (1971) 231; Grub. Opred. rast. Mong. [Key to Plants of Mongolia] (1982) 100. —*S. glauca* With. Bot. arrang. Brit. Pl., ed. 3 (1796) 420; Ser. in DC. Prodr. 1 (1824) 397; Bge. in Ledeb. Fl. alt. 2 (1830) 157; Fenzl in Ledeb. Fl. Ross. 1 (1842) 389; Edgew. et Hook. f. in Hook. f. Fl. Brit. India, 1 (1875) 233; Duthie in Rep. Pamir bound, commiss. (1898) 20; Hanelt und Davazamc in Feddes Repert. 70, 1–3 (1965) 25; Zhou Li-hou in Fl. Xizang. 1 (1983) 700; Claves pl. Xinjiang. 2 (1983) 249. —*S. graminea* var. *glauca* Trautv. in Bull. Soc. natur. Moscou, 32, 1 (1860) 160; Maxim. Enum. pl. Mong. 1 (1889) 101. —*S. filicaulis* Makino in Bot. Mag. Tokyo, 15 (1901) 113; Fl. desert. Sin. 1 (1985) 453. —**Ic.:** Fl. Xizang. 1, tab. 212, fig. 6– 10; Fl. desert. Sin. 1, tab. 165, fig. 12–15.

Described from Europe. Type in Gottingen (GOET) (?).

On mountain meadows, river valleys, meadows, up to 4600 m.

IA. **Mongolia:** *Mong. Alt.* (Grub., l.c.).
IIA. **Junggar:** *Tarb.* (Kobuk river valley, larch forest along brook, July 20, 1914—Sap.).
IIIB. **Tibet:** *South.* ("Lhasa, Nan'mulin', Tszyantszy, Zhen'bu"—Zhou Li-hou, l.c.).
General distribution: Fore Balkh., Jung.-Tarb., Nor. Tien Shan, East. Pam.; Arct., Europe, Mediterr., Balk.-Asia Minor, Fore Asia, Caucasus, Mid. Asia, West. Sib., East. Sib., Nor. Mong., China, Himalayas, Japan.

Note. Local populations transiting into *S. dilleniana* (with relatively small flowers and petals as long as sepals) are often found in habitats not typical for this species.

23. **S. peduncularis** Bge. in Ledeb. Fl. alt. 2 (1830) 157; Turcz. in Bull. Soc. natur. Moscou, 15 (1842) 605; Schischk. in Fl. SSSR, 6 (1936) 411; Grub. Konsp. fl. MNR [Conspectus of Flora of People's Republic of Mongolia] (1955) 127; Opred. rast. Sr. Azii [Key to Plants of Mid. Asia] 2 (1971) 231. — *S. longipes* β *peduncularis* Fenzl in Ledeb. Fl. Ross. 1 (1842) 387; Kryl. Fl. Zap. Sib. 5 (1931) 996. —*S. longipes* auct. non Goldi: Maxim. Enum. pl. Mong. (1889) 101. —**Ic.:** Jc. pl. Fl. Ross. 5, tab. 421; Fl. SSSR, 6, Plate 21, fig. 4.

Described from West. Siberia (Altay). Type in St.-Petersburg (LE).

On coastal meadows, grassy sections on slopes, forests.

IA. **Mongolia:** *Khobd.* (Dzusylan, in forest, July 13, 1879; along south, source of Kharkira river, July 24, 1879—Pot.), *Mong. Alt.* (upper Bulugun river, Ioltin-gola valley near Kudzhurt settlement, Artelin-sala creek valley on south. slope of Shara-Khamryn-sai mountain range, larch forest, July 3, 1971; upper Kobdo river, Dayan-nur, nor. slope of Yamatyn-ula, 2350—2500 m alt., larch forest, July 9, 1971—Grub., Ulzij. et al).
IIA. **Junggar:** *Tien Shan* (Danyu, wet sites in forest, No. 1457, July 17; Chzhaosu area, between Ven'tsyuan' town and Syat, No. 3558, Aug. 15; Kelisu, Daban', on slope, No. 1940, July 18, 1957—Kuan).
General distribution: Arct., West. Sib., East. Sib., Far East, Nor. Mong. (Fore Hubs., Hent., Hang.).

24. **S. pulvinata** Grub. in Bot. zh. 57, 12 (1972) 1592; id. Opred. rast. Mong. [Key to Plants of Mongolia] (1982) 100. —**Ic.:** Grub. Opred. rast. Mong. [Key to Plants of Mongolia] Plate 42, fig. 197.

Described from Mongolia. Type in St.-Petersburg (LE). Map 1.

On rocky slopes and peaks in upper belt, rubble placers, waste lands.

IA. **Mongolia:** *Mong. Alt.* (pass from Chigirtey to Bzau-kul', alpine tundra, July 22, 1908—Sap.; Tolboin-Kungei-nuru mountain range, flat peak of mountain range, alpine belt, fine rubble placers, July 5, 1945—Yun.; Kobdo river basin, Duro-nur lake, Mansar-daba crossing on road to Delyun, 2766 m alt., wormwood-sheep's fescue steppe with, fine cushion structure at outcrops of host rocks, June 30, 1971; Buyantu and Bulugun water divide, Akhuntyn-daba on road to Delyun-Kudzhurtu, 3050 m alt., sedge-*Cobresia* waste land, July 2, 1971; Khadzhingiin-nuru (south of Tsastu-Bogdo), Seterkhi-Khutul', 2844 m alt., pillow-like steppe, June 25, 1971—typus; Ikhe-Ulan-daba crossing on road from Must somon to Uenchi somon, 2945 m alt., on saddle and gentle slope exposed eastward, fine grass sheep's fescue steppe with fine cushion structures; same site, Baga-Ulan-daba, 2845 m alt., saddle, blue grass—sheep's fescue steppe with fine cushion structures, Aug. 13, 1979—Grub., Dariima et al; Baga-Ulan-daba crossing on

road from Must somon to Uenchi, south. slope, on sandy-rocky soil, June 25, 1973, Golubkova and Tsogt).
General distribution: endemic.

25. **S. pusilla** Schmid in Feddes Repert. 31 (1933) 41.
Described from West. Tibet. Type in Zurich (Z). Plate IV, fig. 4.
Habitat not known.

IIIB. Tibet: *South.* (Keptung-La, July 29, Tschusang-po, 5450 m, Aug. 13, Pangong Tso, July 25, 1927 (typus) - Bosshard).
General distribution: endemic.

26. **S. radians** L. Sp. pl. (1753) 422; Ser. in DC. Prodr. 1 (1824) 30; Fenzl in Ledeb. Fl. Ross. 1 (1842) 378; Turcz. in Bull. Soc. natur. Moscou, 15, 3 (1842) 608; Hemsl. in J. Linn. Soc. (London) Bot. 36 (1904) 523; Kryl. Fl. Zap. Sib. 5 (1931) 991; Schischk. in Fl. SSSR, 6 (1936) 418; Grub. Opred. rast. Mong. [Key to Plants of Mongolia] (1982) 100. —*Fimbripetalum radians* Iconnik. (Ikonnik.) in Novit. syst. pl. vasc. 14 (1977) 79. —**Ic.:** Fl. SSSR, 6, Plate 20, fig. 3.
Described from East. Siberia. Type in London (Linn.).
In scrubs.

IA. Mongolia: *East. Mong.* (around Khailar town, in scrubs at foot of sand knolls, June 20, 1951—S.H. Li et al; Khalkhin-gol river valley 35 km beyond Sumber settlement, 720–750 m above mean sea level, July 19, 1987—Dariima et al.).
General distribution: East. Sib., Far East, Nor. Mong. (Fore Hing.), China (Nor.), Japan.

S. schugnanica Schischk. in Fl. URSS, 6 (1936) 882; Ikonnik. Opred. rast. Pamira [Key to Plants of Pamir] (1963) 279; Fl. Tadzh. 3 (1968) 479. —*Mesostemma schugnanica* (Schischk.) Iconnik (Ikonnik.) in Novit. syst. pl. vasc. 13 (1976) 115.
Described from Mid. Asia. Type in St.-Petersburg (LE).
On talus and riverine pebble beds, in upper belt of mountains.

General distribution: Tien Shan, East. Pam.

Note. This species, in all probability, should be found in west. Kashgar and in Chinese Pamir.

27. **S. soongorica** Roshev. in Fl. URSS, 6 (1936) 881; Fl. Tadzh. 3 (1968) 480; Fl. Kazakhst. 3 (1960) 336; Opred. rast. Sr. Azii [Key to Plants of Mid. Asia] 2 (1971) 231; Claves pl. Xinjang. 2 (1983) 248. —**Ic.:** Fl. Kazakhst. 3, Plate 31, fig. 8.
Described from East. Kazakhstan. Type in St.-Petersburg (LE).
In forests, rubble slopes, along rivers, on alpine grasslands.

IB. Kashgar: *Nor.* (Uch-Turfan, June 18, 1908; Ulug-Tuz in Charlym river, among junipers, June 27, 1909—Divn.).

IIA. **Junggar:** *Tien Shan* (in Sary-Kungei mountains, Kadzdy, Aug. 30, 1877; Khanakhai mountains, June 16; Chubaty crossing (Sairam), Aug. 2; in Dzhagastai mountains, Aug. 2; Yugantas, June 25, 1978; near Borborogussun river, June 15; Borgaty crossing, June 7; south. descent from Karagol pass to Nilka, June 16; Kumbel', May 31; on way to Sharysu river, July 26, 1879—A. Per.; Baymkol Tal, am Waldlager an der Mundung des Alai aigut, Sept. 15–18; Klukontk Tal baim Lager unter Tschon Yailak Pass, June 15, 1908—Merzb.; Urumchi region, upper Tasenku river, Biangou site, spruce forest, Sept. 24, 1929—Pop.; nor. slopes of Boro-khoro mountain range, left bank of Yantszykhai, 3000 m alt., in rock crevices, June 23, 1953—Mois.; nor. Yakou, No. 1938, Aug. 31, 1957—Kuan).

General distribution: Jung.-Tarb., cent. Tien Shan; Nor. Mong. (Mong.-Daur.).

Note. Roshevits contrasted this species with *S. brachypetala* Bunge, from which it differs distinctly. *S. soongorica* stands much closer to *S. dahurica* Willd. There are quite a number of cases when difficulties arise in differentiating between these 2 species although they are quite distinctly isolated.

28. **S. strongylosepalata** Hand.-Mazz. in Osterr. Bot. Zeitschr. 88, 4 (1939) 301.

Described from Mongolia. Type in Vienna (W).

Habitat not known.

IA. **Mongolia:** *East.* ("Sain-nor am Fus des Schara-Narin-ula, July 17, 1937, Licent"— Hand.-Mazz. (Mzt.), l.c.—typus).

General distribution: endemic.

S. tibetica Kurz. in Regensb. Fl. (1872) 285; Edgew. et Hook. f. in Hook. f. Fl. Brit. India, 1 (1875) 231; Zhou Li-hou in Fl. Xizang. 1 (1983) 693.

Described from Himalayas (Kashmir). Type in London (K).

On limestone rocks.

Note. This species is known so far only by type specimens collected "on limestone rocks near Trantse Lundo in Karnag, 4200–5200 m alt., Stoliczka". Probably, it is also found in Tibet but floristic investigations there are not adequate so far. The existing citation for west. Tibet (3600–5500 m alt.— Zhou Li-hou, l.c.) probably pertains to Himalayas.

S. turkestanica Schischk. in Tr. Bot. muz. Ak. nauk, 24 (1932) 31; Schischk. in Fl. SSSR, 6 (1936) 401; Ikonnik. Opred. rast. Pamira [Key to Plants of Pamir] (1963) 104; Fl. Tadzh. 3 (1968) 478; Opred rast. Sr. Azii [Key to Plants of Mid. Asia] 2 (1971) 229. —**Ic.:** Fl. SSSR, 6, Plate 21, fig. 8; Fl. Tadzh. 3, Plate 77, fig. 3, 4.

Described from Mid. Asia (Zeravshan). Type in St.-Petersburg (LE).

On rocky slopes and talus in upper belt of mountains.

General distribution: East. Pam.; Mid. Asia: West. Tien Shan.

Note. This species, in all probability, should be found in west. Kashgar and in Chinese Pamir.

29. **S. viridescens** (Maxim.) Ju. Kozhevn. in Novit. syst. pl. vasc. 20 (1983) 106. —*S. graminea* var. *viridescens* Maxim. Fl. Tangut. 1 (1889) 90; id. in Acta Horti Petrop. 11, 1 (1889) 72.

Described from Qinghai. Type in St.-Petersburg (LE). Plate III, fig. 2.
On shoals, rocks.

IIIA. Qinghai: *Nanshan* (South Kukunor mountain range, on rocks, 3150–3500 m alt., June 7, 1880—Przew.; on Itel'-gol river, July 1885—Pot.; south. bank of Tetung river, July 11, 1872—Przew., typus !).
General distribution: China (Nor.-West.).

S. williamsiana Ju. Kozhevn. nom. nov. —*Arenaria monosperma* Will. in J. Linn. Soc. (London) Bot. 38 (1907–1909) 398. —*S. monosperma* (Will.) Ju. Kozhevn. in Novit. syst. pl. vasc. 20 (1983) 106, non Buch.-Ham. in D. Don. (1825).

Described from Himalayas. Type in London (K).
Habitat not known.

Note. The species has been described on the basis of King's collection (1882) in Tibet as shown on the label. However, the territory regarded as Tibet by King in fact does not fall in Tibet at all.

The description of this species by F. Williams (l.c.) within genus *Arenaria* is surprising since the 2 type specimens clearly reveal that petals are deeply laciniated and the capsule dehisces on 6 valves (reported by Williams as well). Some capsules contain many seeds, instead of 1, and hence the tag "monospermy" is of no taxonomic significance.

The species stands closest to *S. graminea* L.

30. **S. winkleri** (Briq.) Schischk. in Fl. SSSR, 6 (1936) 403; Ikonnik. Opred. rast. Pamira [Key to Plants of Pamir] (1963) 104; Fl. Tadzh. 3 (1968) 478; Opred. rast. Sr. Azii [Key to Plants of Mid. Asia] 2. (1971) 230; Podlech, Angers in Mitt. Bot. Staatssaml. München, 13 (1977) 418. —*Cerastium winkleri* Briq. in Ann. Conserv. et Jard. Bot. Geneve, 13–14 (1911) 382. — *Cerastium schizopetalum* H. Winkl. in Vidensk. Meddel. (1901) 51, non Maxim. —Ic.: Fl. Tadzh. 3, Plate 78, fig. 5.

Described from .east. Pamir. Type in Copenhagen (C). Plate 1, fig. 2.
On moist sites in alpine belt.

IC. Qaidam: *Plain* (Ikhe-Tsaidamin-nor lake, June 14, 1985—Rob.).
IIIA. Qinghai: *Nanshan* (Sharagol'dzhin river, Baga-bulak area, June 15, 1894—Rob.).
General distribution: East. Pam.

Note. Thus, there is disjunction in the longitudinal distribution of this species almost throughout the whole of Central Asia.

2. **Pseudostellaria** Pax in Engl. und Prantl

Naturl Pfl. fam. ed. 2, 16c (1934) 318. —*Krascheninnikovia* Turcz. ex Fenzl in Endl. Gen. pl. (1840) 968; Schaeftlein in Taxon, 6, 4 (1957) 139.

1. Seedcoat covered with 3–4-radial anchorlike tubercles or bristles. Leaves soft, tender. Tubers up to 0.8 cm long, turnip-shaped. Stems usually weak, lodging, often forming loose tufts. ....
.... 2. **P. rupestris** (Turcz.) Pax.
+   Seedcoat covered with tubercles with a bristle on top of tubercle. Leaves resilient. Tubers up to 1–2 cm long, carrot-shaped. Stems usually strong, erect, solitary.. .... 1. **P. heterantha** (Maxim.) Pax.

1. **P. heterantha** (Maxim.) Pax in Engl. und Prantl, Naturl. Pfl. fam. ed. 2, 16c (1934) 318; Ohwi in J. Jap. Bot., 9, 1 (1937) 101. —*Krascheninnikovia heterantha* Maxim. in Mel. biol. 9 (1873) 38; id. in Acta Horti Petrop. 9 (1890) 71; Franch. et Savat. Enum. pl. Jap. 2 (1879) 297; Korsh. in Bull. Ac. Sci. St.-Petersb. 5, 9 (1898) 40; Koidz. Flor. symb. or.-as. (1930) 24. — *Krascheninnikowia maximowicziana* Franch. et Savat. Enum. Pl. Jap. 2 (1879) 297; Maxim. Fl. Tangut. (1889) 85; id. in Acta Horti Petrop. 9 (1890) 70, p.p.; Korsh. l.c.; Murav'eva in Fl. SSSR, 6 (1936) 429. —*Stellaria davidii* var. *himalaica* Franch. in Bull. Soc. Bot. France, 33 (1886) 434. —*Krascheninnikovia himalaica* (Franch.) Korsch. in Bull. Ac. Sci. St.-Petersb. 5, 9 (1898) 40; Williams in J. Linn. Soc. (London) Bot. 38 (1907–1909) 397, p.p.; Hand.-Mazz., Symb. Sin. 7 (1929) 193. —*Stellaria heterantha* (Maxim.) Franch. Pl. Delav. (1890) 101. —*Pseudostellaria maximowicziana* (Franch. et Savat.) Pax, l.c., p.p.; Zhao Li-hao in Acta Sci. natur. univ. Intramong. 19, 4 (1988) 672. —*P. himalaica* (Franch.) Pax, l.c. p.p. —*P. heterantha* var. *himalaica* Ohwi, l.c.; Zhou Li-hou in Fl. Xizang. 1 (1983) 706. —**Ic.**: Fl. SSSR 6, Plate 22, fig. 4.

Described from Japan (vicinity of Nagasaki). Type in St.-Petersburg (LE). In mountainous coniferous forests and in alpine belt.

IIIA. Qinghai: Amdo (in Mudzhik mountains, June 19–21; in alpine belt of Dzhakhar-Dzhargyn mountains, in juniper forests, Feb. 5, 1880—Przew.).
IIIB. Tibet: Weitzan (Dzhagyn-gol river, on river bank, 4050 m, July 3, 1900—Lad.).
General distribution: Far East, China (Nor., Nor.-West.), Himalayas (east.), Japan.

Note. Plants with short-ciliate leaf margin belong to *P. tibetica* Ohwi. Ohwi (1937) pointed out that his new species differs from *P. heterantha* in broader leaves, short-ciliate all along their margin. On the whole, these species are similar. However, leaf breadth in *P. heterantha* is highly inconstant since generally this species is polymorphous. Ciliate leaf margin can serve as a distinguishing characteristic for the variety, but not species. The following combination is therefore necessary: *P. heterantha* (Maxim.) Pax var. *tibetica* (Ohwi) Ju. Kozhevn. —*P. tibetica* Ohwi in J. Jap. bot. 9, 1 (1937) 103; Zhou Li-hou in Fl. Xizang. 1 (1983) 705, Plate 226, fig. 1–5.

2. **P. rupestris** (Turcz.) Pax in Engl. und Prantl, Naturl. Pfl. fam. ed. 2, 16c (1934) 318; Ohwi in J. Jap. bot. 9, 1 (1937) 98. —*Krascheninnikowia rupestris* G. et T. ex Besser in Flora, 17, 1 Beibl. (1834) 9; Turcz. ex Fenzl in Endl. Gen. pl. (1840) 968; id. in Bull. Soc. natur. Moscou, 15 (1842) 609; Fenzl in Ledeb. Fl. Ross. 1 (1842) 373; Regel, Pl. Raddeanae, 1, 9 (1862) 379; Maxim. in Mel. biol. 9 (1873) 37; Korsh. in Bull. Ac. Sci. St.-Petersb. 5, 9 (1898) 40; Murav'eva in Fl. SSSR, 6 (1936) 426; Grub. Konsp. fl. MNR [Conspectus of Flora of People's Republic of Mongolia] (1955) 127; id. Opred. rast. Mong. [Key to Plants of Mongolia] (1982) 101; Zhao in Acta sci. natur. univ. Intramong. 19, 4 (1988) 671. —*Stellaria rupestris* (Turcz.) Hemsl. in Forbes and Hemsl. in J. Linn. Soc. (London) Bot. 23 (1886) 69 (non Scop.); Franch. Pl. Delav. (1890) 101. —*Krascheninnikowia borodinii* Kryl. in Mel. bot. Borod. (1927) 220. —Ic.: Fl. SSSR, 6, Plate 22, fig. 6; Grub. Opred. rast. Mong. [Key to Plants of Mongolia] Plate 17, fig. 195.

Described from East. Siberia (Transbaikal). Type in Kiev (KW).

In larch forests, on rocks, among stones on slopes and peaks of mountains.

IA. Mongolia: *Mong. Alt.* (Khara-Adzarga mountain range, around Khairkhan-Duru river, in larch forest, Aug. 26, 1930—Pob.).

IIIB. Tibet: *Weitzan* (Dzhagyn-gol river, No. 177, 1900—Lad.).

General distribution: West. Sib. (Altay), Far East, Nor. Mong. (Hent., Hang., Mong.-Daur.), Japan.

3. **Myosoton** Moench

Meth. pl. (1794) 225 —*Malachia* Fries, Fl. Hall. (1818) 77 —*Stellaria* subgen. *Myosoton* (Moench) Pax.

1. **M. aquaticum** (L.) Moench, l.c. —*Cerastium aquaticum* L. Sp. pl. (1753) 439; Ledeb. Fl. alt. 2 (1830) 182. —*Malachia* (*Malachium*) *aquatica* Fries, l.c.; Kar. et Kir. in Bull. Soc. natur. Moscou, 15, 1 (1842) 174; Trautv. in Bull. Soc. natur. Moscou, 32, 1 (1860) 162; Murav'eva in Fl. SSSR, 6 (1936) 430; Opred. rast. Sr. Azii [Key to Plants of Mid. Asia] 2 (1971) 233; Zhou Li-hou in Fl. Xizang. 1 (1983) 662. —*Stellaria aquatica* (L.) Scop. Fl. Carniol. 1 (1772) 319; Edgew. in Hook. f. Fl. Brit. India, 1 (1875) 229; Forbes and Hemsl. in J. Linn. Soc. (London) Bot. 23 (1886) 67; Hand.-Mazz. Symb. Sin. 7 (1929) 188; Hao in Bot. Jahrb. 68 (1938) 594.

Described from Europe. Type in London (Linn.).

On moist banks of rivers and gorges, up to 3700 m.

IIA. Junggar: *Tarb.* (flussufer Tschurtschutsu bei Tschugutschak, Aug. 10, 1840—Schrenk).

IIIB. Tibet: *South.* ("in Lhasa region"—Zhou Li-hou, l.c.).

General distribution: Europe, Mediterr., Balk.-Asia Minor, Caucasus, Mid. Asia, West. Sib., China (Nor., Nor.-West., Nor.-East., Cent., South., Taiwan), Himalayas, Korea, Japan.

4. Cerastium L.

Sp. pl. (1753) 437.

1. Styles 2-3 (rarely 4). Capsule dehiscent on 6 teeth. Hairs with purple septa sometimes present in pubescence ....
    ....3. **C. cerastoides** Britt.
+   Styles 5. Capsule dehiscent on 10 teeth. Hairs with purple septa absent in pubescence ....2.
2. Teeth of mature capsule revoluted downward externally. Sepals ovoid, short-cuspidate, up to 3 mm or more broad .... 3.
+   Teeth of mature capsule erect with margins recurved externally. Sepals lanceolate, long-cuspidate, not more than 2.5 mm broad....7.
3. Sepals glabrous; rest of plant parts with only rare cilia. Leaf bases amplexicaul. Rhizome with thickenings ....
    .... 4. **C. davuricum** Fisch. ex Spreng.
+   Sepals pubescent; rest of plant parts also more or less pubescent. Leaf bases not amplexicaul. Rhizome without thickenings .... 4.
4. Pubescence of sepals with simple eglandular hairs .... 5.
+   Pubescence of sepals with glandular and eglandular hairs ...... 6.
5. Sepals pubescent predominantly in lower part. Flowers on short peduncles in terminal inflorescences numbering 3 to 7. Plant up to 30 cm tall, green, sparsely pubescent. Leaves linear or lanceolate.
    .... .5. **C. falcatum** Bge.
+   Sepals pubescent all over surface. Flowers on long peduncles, usually solitary (rarely 2–3). Plant up to 10 (rarely 15) cm tall, glaucescent due to compact pubescence. Leaves oval. ....
    .... .6. **C. lithospermifolium** Fisch.
6. Sepals 4–6 mm long. Petals crenate or crispate (not emarginate) on margin. Leaves ovoid, rather short-cuspidate, with breadth to length ratio 1:3; long-ciliate along margin.. ....
    .... 8. **C. pauciflorum** Stev.
+   Sepals 6–10 mm long. Petals emarginnate for 1/4 (rarely 1/3). Leaves lanceolate, long-cuspidate, with breadth to length ratio 1:6; short-ciliate along margin ....7. **O. maximum** L.
7. Annual plants. Sepals 4–5 mm long .... 8.
+   Perennial plants. Sepals 5–9 mm long... 9.
8. Leaves oval, 4–10 mm broad. Multicellular hairs abundant in pubescence. Plant up to 35 cm tall. Peduncles as long as calyx or shorter, not deflexed; inflorescence agglomerated ....
    .... **C. glomeratum** Thuill.
+   Leaves lanceolate, 2–3 mm broad. Pubescence with only unicellular hairs. Plant up to 15 cm tall. Peduncles somewhat

longer than calyx, deflexed; inflorescence lax ....
....9. **C. pumilum** Curt.
9. Axils of cauline leaves with reduced shoots ... .2. **C. arvense** L.
+ Reduced shoots in axils of leaves absent.................................... 10.
10. Petals twice as long as calyx ....1. **C. alpinum** L.
+ Petals shorter, equal to or somewhat longer than calyx............ 11.
11. Leaves ovoid, 5–8 mm broad, long-cuspidate. Plant up to 30 cm
tall .... 11. **C. vulgatum** L.
+ Leaves oval-oblong, 2–3 mm broad, short-cuspidate or rounded.
Plant up to 15 cm tall .... 10. **C. pusillum** Ser.

1. **C. alpinum** L. Sp. pl. (1753) 438; Ser. in DC. Prodr. 1 (1824) 419; Ledeb.
Fl. alt. 2 (1830) 180; Fenzl in Ledeb. Fl. Ross. 1 (1842) 412, cum α *hirsutum*,
β *lanatum*, γ *glabratum*; Trautv. in Bull. Soc. natur. Moscou, 32, 1 (1860) 161;
Maxim. Enum. pl. Mong. (1889) 105; id. Fl. Tangut. (1889) 93; Hemsl. in J.
Linn. Soc. (London) Bot. 36 (1904) 464; O. Fedtsch. Rast. Pamira [Plants of
Pamir] (1904) 19; Marquand in J. Linn. Soc. (London) Bot. 48 (1929) 165;
Schischk. in Fl. SSSR, 6 (1936) 458. —*C. fontanum* var. *tibetica* (Edgew. et
Hook. f.) C.Y. Wu et L.H. Zhou in Fl. Xizang. 1 (1983) 663. —*C. vulgatum*
var. *tibetica* Edgew. et Hook. f. in Fl. Brit. India, 1 (1874) 228. —**Ic.:** Fl.
Xizang. tab. 214, fig. 1–6.

Described from Europe. Type in London (Linn.). Plate V, fig. 1.

On meadowy and rubbly slopes in alpine belt, banks of brooks, rock
screes and talus.

IA. Mongolia: *Mong. Alt.* (Mal. Kairta—Mal. Ku-Irtys crossing, July 17, 1908—
Sap.; 35 km south-west of Khalyun settlement, under shade of rocks in forest belt,
Aug. 14, 1984—Gub.).

IIA. Junggar: *Tien Shan* (Narat crossing, June 15, 1877—Przew.; Urten-Muzart
crossing, Aug. 2, 1878—Fet.; Kumbel', May 30, 1879—A. Reg.).

General distribution: Jung.-Tarb., Nor. and Cent. Tien Shan, East. Pam.; Arct.
(Europ.), Europe, Nor. Mong., China (Nor., Nor.-West.), Himalayas, Japan, Nor. America.

Note. This species differs very little from *C. vulgatum* L. and it would
probably be more appropriate to assign it the rank of variety of latter species.

2. **C. arvense** L. Sp. pl. (1753) 438; Ser. in DC Prodr. (1824) 419; Fenzl in
Ledeb. Fl. Ross. 1 (1842) 412; Trautv. in Bull. Soc. natur. Moscou, 32, 1
(1860) 162; Forbes and Hemsl. in J. Linn. Soc. (London) Bot. 23 (1886) 66;
Maxim. Enum. pl. Mong. (1889) 105; Kryl. Fl. Zap. Sib. 5 (1931) 1015;
Schischk. in Fl. SSSR, 6 (1936) 460; id. in Fl. Zabaik. 4 (1941) 308; Grub.
Konsp. fl. MNR [Conspectus of Flora of People's Republic of Mongolia]
(1955) 128; Opred. rast. Sr.Azii [Key to Plants of Mid. Asia] 2 (1971) 238;
Grub. Opred. rast. Mong. [Key to Plants of Mongolia] (1982) 101; Ma Yu-
chuan in Fl. Intramong. 2 (1978) 178, cum var. *angustifolium* et var. *glabellum*

(Turcz.) Fenzl; Claves pl. Xinjiang. 2 (1983) 240; Fl. desert. Sin. 1 (1985) 449, pro var. *angustifolium* Fenzl. —*C. incanum* Ledeb. in Mem. Ac. Sci. St.-Petersb. 5 (1835) 540; Turcz. in Bull. Soc. natur. Moscou, 15, 3 (1842) 614. — Ic.: Fl. Zabaik. 4, fig. 162; Fl. Intramong. 2, tab. 94, fig. 1–3; Fl. Kazakhst. 3, Plate 32, fig. 1; Fl. desert. Sin. tab. 164, fig. 1–4.

Described from Scandinavia. Type in London (Linn.).

In coastal and alpine meadows, forests and mountain steppes, rocks and talus.

IA. **Mongolia:** *Khobd.* (east. fringe of Achit-nur lake, Ulyasutuin-gol river valley, Aug. 1947—Tarasov), *Mong. Alt.* (upper Indertiin-gol, subalpine steppe, July 25; alps at Dolon-nor, on granite rocks, July 8, 1877—Pot.; Dain-gol lake, June 28, 1903; Mal. Kairta—Mal. Ku-Irtis crossing, on talus, July 17, 1908—Gr.-Grzh.; nor. slope of Khara-gol mountain range, larch forest along nor. slope of mountain, Aug. 17, 1930—Pob.; Taishiri-ula mountain range, nor., slope, larch forest, July 12, 1945—Yun.; nor. slope of Khan-Taishiri mountain range, patch of larch forest 15 km south-east of Yusun-Bulak, under cover, Sept. 1, 1948—Grub.; same site, south. slope in upper Shine-usu river, 2380–2450 m, steppe, June 19, 1971—Grub., Ulzij. et al; upper Kobdo river, Dayan-nur, nor. slope of Yamatyn-ula, 2350–2500 m, larch forest, July 9; same site, Khoton-gola valley, east. slope of Madat-Tologoi town, near winter camp, 2400 m, July 8—1971; Khasagtu-Khairkhan, nor. slope of Tsagan-Irmyk-ula, upper Khunkerin-ama, *Cobresia* grove on nor. slope, 2700–3100 m, Aug. 23, 1972—Grub., Dariima et al; *Shazdgat-nuru mountain range, Ulyastei-gola valley, 2100 m, in *Cobresia* meadows, June 27, 1973—Golubkova, Tsogt), *East. Mong.* (*between Kulusutaevsk and Dolon-Nor, 1870—Lom.; Suma-Khada mountain range, June 12, 1871—Przew.; *Khailar railway station, July 10, 1902—Litw.; Arshan town, mountain meadow, June 13, 1950—Yu-l. Chang; Khailar town, Sishan' hillocks, July 7, 1951—Skvortzov et al; Khailar town, Nunlin'tun' village, in scrubs on sandy slope, June 10, 1951—Wang; *Khamar-daba region, floodplain, Khalkh-gol river, June 19, 1954—Dashnyam; Shiliin-Bogdo-Ula, nor. slope, about 1650 m, No. 9791, July 12, 1985—Gub.), *Depr. Lakes* (Kharkira river valley in Ulangom region, peaks of small mountains, Sept. 29, 1931—Baranov, Shukhardin), *Gobi. Alt.* (Dundu-Saikhan mountains, south-west. slope of upper belt, July 8; same site, slopes of upper belt, in moist sites, July 9, 1909—Czet.; "nor. part of Dzun-Saikhan mountain range, upper 1/3 of nor. slope, on bank, June 19, 1945—Yun.), *Alash. Gobi* (Alashan mountain range, Yamata gorge, slopes of upper belt, June 13, 1907–1909, Czet.).

IIA. **Junggar:** *Cis-Alt.* (Korumduk mountain around Kurtu river, June 16, 1903—Gr.-Grzh.; Koktogai-Fuven', No. 1927, Aug. 17, 1956—Ching), *Tarb.* (north of Dachen [Chuguchak] No. 1583, Aug. 13, 1957—Kuan), *Jung. Alat.* (Toll district, on slope, No. 2639, Aug. 6, 1957—Kuan), *Tien Shan* (*Nan'shan-kou, near spring, May 26, 1877—Pot.; Sairam, south-east. bank, July 22, 1877—A. Reg.; Karlyk-taga slope, Sept. end; Tushirtyn-gol mountain range, under rocks, Aug. end, 1895; mountains near Santash pass, spruce forest, June 18, 1893—Rob.; Barkul' lake, Nanshan, in mountain gorge, No. 4947, Sept. 27, 1957—Kuan; in Arashan, Mal. Yuldus, on valley proluvium, No. 6302, Aug. 2, 1958—S.H. Li et al).

General distribution: panboreal, South Amer.

Note. The following varieties are distinguished within the species: var. *angustifolia* Fenzl with predominantly 1.5-2.5 mm broad leaves and var. *latifolia* Fenzl with predominantly 2–4 mm broad leaves (asterisked*). Plants totally devoid of pubescence were collected in east. Mongolia.

3. **C. cerastoides** (L.) Britt. in Mem. Torrey Bot. Club, 5 (1894) 150; Schischk. in Fl. SSSR, 6 (1936) 435; Grub. Konsp. fl. MNR [Conspectus of Flora of Mongolian People's Republic] (1955) 128; id. Opred. rast. Mong. [Key to Plants of Mongolia] (1982) 101; Ikonnik. Opred. rast. Pamira [Key to Plants of Pamir] (1963) 106; Opred. rast. Sr. Azii [Key to Plants of Mid. Aaia] 2 (1971) 234; Podetch. Angers in Mitt. Bot. Staatssaml. München, 13 (1977) 415; Claves pl. Xinjiang. 2 (1983) 239. —*Stellaria cerastoides* L. Sp. pl. (1753) 422; Ledeb. Fl. alt. 2 (1830) 155; Turcz. in Bull. Soc. natur. Moscou, 15 (1842) 597. —*Cerastium trigynum* Vill. Hiat. pl. Dauph. 3 (1789) 645; Fenzl in Ledeb. Fl. Ross. 1 (1842) 396, cum α *grandiflorum*, β *glandulosum* et γ *parviflorum;* Trautv. in Bull. Soc. natur. Moscou, 32, 1 (1860) 160; Boiss. Fl. or. 1 (1867) 715; Edgew. et Hook. f. in Fl. Brit. India, 1 (1874) 227; O. Fedtsch Rast. Pamira [Plants of Pamir] (1904) 19; Hedin, S. Tibet (1922) 84. —*C. obtusifolium* Kar. et Kir. in Bull. Soc. natur. Moscou, 14, 3 (1841) 393; Fenzl in Ledeb. Fl. Ross. 1 (1842) 398. —*Dichodon cerastoides* (L.) Reichb. Ic. Fl. Germ. 5 (1842) 34; Ikonnik. Opred. vyssh. rast. Badakhsh. [Key to Higher Plants of Badakhsh (1979) 147. —**Ic.**: Fl. SSSR, 6, Plate 12, fig. 8.

Described from Scandinavia. Type in London (Linn.).

In forests, on alpine and subalpine meadows, along rivers, on rocks.

IA. Mongolia: *Khobd.* (Dzusylan, along lower border of larch forest, July 11, 1879—Pot.), *Depr. Lakes* (*Dzun-Dzhirgalantu mountain range, south-west. slope, Ulyastyn-gola gorge, 1850–2800 m, June 28, 1971—Grub., Ulzij. et al), *Mong. Alt.* (Daban, toward Dzasaktu-khan, July 12, 1877—Pot.; Khara-Adzarga mountain range, around Khairkhan-duru, larch forest, Aug. 26, 1930—Pob.; Adzhi-Bogdo mountain range, Burgasin-daba pass between Indertiin-gol and Dzuslangin-gol, rubble placers of alpine belt, Aug. 6, 1947—Yun.; *near Tonkhil somon, among rocks, June 24, 1975—Golubkova and Tsogt; Tsastu-Bogdo-ula in upper Dzuilin-gol, south-east. slope, 2950–3000 m, on pillow type formations, June 24, 1971—Grub., Ulzij. et al; upper Indertiin-gol river, marshy meadow in alpine belt, July 24; Kharagait-Khutul' pass, near lower boundary of forest, July 27, 1947—Yun.).

IIA. Junggar: *Cis-Alt.* (larch forest on west. side of Kandagatai river valley, Sept. 18, 1876—Pot.; Khobuk river, larch forest on rivulet, July 20, 1914—Sap.; Qinhe-Khun'tai and Qinhe, in steppe, No. 967, 1078, Aug. 4, 1956—Ching; in Altay mountains, on slope, July 17; Koktogai, alongside of river, June 6, 1959—Lee and Chu), *Tarb.* (on Ui-Chilik river, Sept. 21, 1876—Pot.; south. slope of Saur mountain range, Karagaitu river valley, subalpine meadow, June 23, 1957—Yun. et al; Dachen-Tarbagatai, No. 1659, Aug. 14; north of Dachen, No. 1573, Aug. 13—1957, Kuan), *Jung. Alat.* (Syata-Ven'tsyuan', No. 1433, Aug. 13; in Toli district, No. 1704, Aug. 6; 30 km west of Ven'tsyuan', in Ala Tau mountains, on water divide, No. 2019, Aug. 25—1957, Kuan), *Tien Shan* (*Kul'dzha, July 12; *Iren-Khabirga, July 28; Karagol, June 17; *Kumbel', May 31—1979, A. Reg.; Passe zwischen Kin und Kurdai, July 3; Mittleres Kurdai-Tal, ende July, 1907; am sudrande des Bogdo-ola, Aug. 28–29, 1908—Merzb.; *Santai, upper part of Shi-gou river valley (or Tshka-su), unflooded erosion terrain, 2000 m, May 27, 1952—Mois.; 10 km nor. of Chzhaosu town, No. 3308, Aug. 15; 15 km south of Tyan'chi, in Fukan district, Sept. 19; south of Yakou, on nor. slope, No. 1957, Aug. 31; in Fukan district, in Tyan'chi lake region, on slope, No. 4327, Sept. 19; Guchen district, in Magolyan, in steppe, Sept. 22; between Nilki and Qinhe, on rock talus, No. 4053, Sept. 1; in Yakou between Nilki and Ulyasutai, on slope, No. 3957, Aug. 30—

1957, Kuan), *Jung. Gobi* (Tsitai-Beidashan', on shaded slope, Sept. 28, 1959—Lee and Chu).

IIIA. Qinghai: *Nanshan* (Mon'yuan, glacial moraine at Ganshig river source, Peishikhe river tributary, Aug. 18, 1958—Dolgushin).

IIIB. Tibet: *Chang Tang* ("Chimen-tagh, kan-yalak-sai, 3984 m, July 21, 1900"—Hedin, 1922).

IIIC. Pamir (*Biluli river near its discharge into Gumbus river, between rocks on brook, June 11, 1909—Divn.; along Mia river gorge, 4000 m, Aug. 21, 1941; same site, upper Kanlyk river, 1500–5000 m, alpine tundras, July 14; same site, Taspestlyk area, 4000–5000 m, July 25, 1942—Serp.).

General distribution: Fore Balkh., Jung.-Tarb., Nor. and Cent. Tien Shan, East. Pam.; Arct. (Europ.), Europe, Balk.-Asia Minor, Fore Asia, Caucasus, Mid. Asia, West. Sib. (Altay), East. Sib. (Sayans), Far East, Nor. Mong., China, Himalayas, Nor. Amer.

Note. The broad-leaved var. *foliosum* Ju. Kozhevn., var. nov., folia obovata, breve acutata, differs from the type variety with linear leaves. Typus: Zinchai, Monjuan. Ad moraenam juxta fontem fl. Ganshinga, affluxio fl. Peischiche, Aug. 18, 1958, L. Dolguschin (LE). Plate III, fig. 4.

The varieties differentiated by Fenzl, in our opinion, are only forms.

4. C. davuricum Fisch. ex Spreng. Pl. minus cognit. 2 (1815) 65; Ser. in DC. Prodr. 1 (1824) 415; Ledeb. Fl. alt. 2 (1830) 177; Gren. Monogr. Cerast. (1841) 13; Fenzl in Ledeb. Fl. Ross. 1 (1842) 401, cum α *glabrum* et β *pilosum*; Kar. et Kir. in Bull. Soc. natur. Moscou, 15, 1 (1842) 174; Turcz. in Bull. Soc. natur. Moscou, 15, 3 (1842) 612; Trautv. in Bull. Soc. natur. Moscou, 32, 1 (1860) 161; Edgew. et Hook. f. in Fl. Brit. India, 1 (1874) 227; Rgl. Descr. pl. 5 (1877) 35; Maxim. Enum. pl. Mong. (1889) 104; Schischk. in Fl. SSSR, 6 (1936) 444; Grub. Konsp. fl. MNR [Conspectus of Flora of People's Republic of Mongolia] (1955) 128; Opred. rast. Sr. Azii [Key to Plants of Mid. Asia] 2 (1974) 235; Grub. Opred. rast. Mong. [Key to Plants of Mongolia] (1982) 101; Claves pl. Xinjiang. 2 (1983) 240. —Ic.: Fl. Kazakhst. 3, plate 32, fig. 7.

Described from East. Siberia (Transbaikal). Type in St.-Petersburg (LE).

In forests, alpine meadows, along river banks.

IIA. Junggar: *Cis-Alt.* (Kandagatai, in ravines, Sept. 15, 1876—Pot.; border of Shara-Sume town, No. 2639, Sept., 3; in Burchum, No. 3084, Sept. 14, 1956—Ching), *Tarb.* (nor.-west of Khob-saira (Chagan-obo mountain), June 22, 1959—Lee and Chu; Dachen, No. 1637, Aug. 13; north of Dachen, No. 1519, Aug. 17; in Dachen region, No. 2934, Aug. 13, 1957—Kuan), *Jung. Alat.* (in Toli district, on subalpine meadows, No. 1049, Aug. 6; Toli district, No. 2591, Aug. 6; mountains around Toli, No. 1157, Aug. 7, 1957—Kuan), *Tien Shan* (Sairam, Aug., Talki gorge, July 18, 1877; Dzhagastai, June 20; on Kassan river, June 22, 1878; Arystyn, July 12; Karagol, June 16, 1879—A. Reg.; between Taldy-bulak and Kegen rivers, June 18; on Sairam lake, July 23; Kutukshi, west of Kul'dzha, June 6, 1878—Fet.; lake at foot of Bogdo-ula, Aug. 29, 1898—Klem.; in mittleren Kukurtuk Tal. geht jedoch hoch hinauf bis 3200 m und hoher, June 22 to July 2; Kapkaktal, feuchte alpenwiesen, July 20, 1903; Syrt von Karabulak und Kara-dschon, July 50, 1907—Merzb.; Urumchi region, upper Tasenku region, Biangou locality, Sept. 24, 1929—Pop.; Urumchi—Nanshan, No. 0554, July 21, 1956—Ching; Manas river basin, left bank, Ulan-Usu river valley, 8-9 km above confluence of Koi-su in it, on pebble bed floor, July 18, 1957—Yun. et al; Sin'yuan',

Nanshan, in forest, No. 1123, Aug. 22; mountains south of Nyutsyuan'-tsza, No. 638, July 18; 15 km south of Tyan'chi lake, No. 1957, Sept. 19; south of Shichan, in Savan district, No. 814, July 23; 8 km south of Nyutsyuan'tsza, on subalpine meadows, No. 679, July 19; Chzhaosu district, mountains in Aksu region, No. 3507, Aug. 14; Sin'yuan', Nanshan, No. 3731, Aug. 22; in Ili-Chapchal, in Dzhagastai, No. 3192, Aug. 8; Kelisu-Karayuz, slightly north of Turfan, No. 1823, July 16; in Savan district, Datszymyao, No. 1711, July 22, same site, on water, No. 1262, July 8, 1957—Kuan).

General distribution: Jung.-Tarb., Nor. Tien Shan; Cent. Tien Shan, Europe, Balk.-Asia Minor, Caucasus, West. Sib. (Altay), East. Sib. (Sayans), Nor. Mong. (Mong.-Daur.), China (Nor.-West.).

Note. The authors used the wrong spelling "dahuricum" at many places.

5. **C. falcatum** Bge. in Mem. Ac. Sci. St.-Petersb. 2 (1835) 549; id. Suppl. Fl. alt. (1836) 37; Fenzl in Ledeb. Fl. Ross. 1 (1842) 398; Kar. et Kir. in Bull. Soc. natur. Moscou, 15, 1 (1842) 175; Trautv. in Bull. Soc. natur. Moscou, 32, 4 (1860) 160; Maxim. Enum. pl. Mong. (1899) 104; Schischk. in Fl. SSSR, 6 (1936) 439; Claves pl. Xinjiang. 2 (1983) 2340; Fl. desert. Sin. 1 (1985) 451. —*C. lithospermifolium* Bge. Fl. alt. 2 (1830) 179, non Fisch. —*C. maximum* β *falcatum* Rgl. Pl. Radd. 1 (1862) No. 348. —*C. bungeanum* Vved. in Fl. Uzb. 2 (1953) 353; Fl. Tadzh. 3 (1968) 484; Opred. rast. Sr. Azii [Key to Plants of Mid. Asia] 2 (1971) 235. —Ic.: Fl. SSSR, 6, Plate 22, fig. 10; Fl. desert. Sin. tab. 164, fig. 8; Fl. Tadzh. 3, Plate 78, fig. 7–9.

Described from Altay. Type in St.-Petersburg (LE).

In meadows, scrubs.

IB. Kashgar: *Nor.* (Uch-Turfan, June 20, 1908—Divn.).

IIA. Junggar: *Cis-Alt.* (Bugotor area, June 11; Korunduk mountain around Kurtu river, June 16—1903, Gr.-Grzh.), *Tien Shan* (Kul'dzha and west of it, May 8; Kuiankuz, nor. of Ili, June 19; Kitentass, April 17, 1877; Agyaz river, July 1878; Sredn. Taldy, May 25; Pilyuchi gorge, April 22; Dzagastai-gol, Sept. 5, 1879—A. Reg.; Kul'dzha, 1876, Golike; same site, 1906, Muromskii; Kungess Tal, May 1–5, 1908—Merzb.; in Aksu valley, July 4, 1878—Fet.; Urumchi dacha, near irrigation canal, May 9, 1954—Mois.; 7 km south-south-east of Urumchi, near Yan'ervo settlement, Urumchinki river valley, along fringes of small silted streams, May 31, 1957—Yun. et al; Urumchi, Yan'ervo settlement, No. 538, May 31, 1957—Kuan; 20 km south-east of Urumchi, No. 6005, May 25; 10 km south-east of Bain-bulak, in Kaidukhe valley, in floodplain, No. 6489, Aug. 11, 1958—Lee and Chu), *Jung. Gobi* (Sochzhan river, June 13, 1877—Pot.; Shankhau-syan Mountain range, near spring, May 24, 1879—Przew.; west. spur of Khuvchiin-nuru mountain, 8 km south-west of Maikhan-Ulan mountains, No. 819, Aug. 1, 1984—Dariima, Kamelin).

General distribution: Fore Balkh., Jung.-Tarb., Tien Shan; West. Sib. (Altay), Mid. Asia.

C. glomeratum Tuill. Fl. Paris, ed. 2 (1799) 225; Schischk. in Fl. SSSR, 6 (1936) 450; Fl. Tadzh. 3 (1968) 488; Opred. rast. Sr. Azii [Key to Plants of Mid. Asia] 2 (1971) 236. —*C. viscosum* L. Sp. pl. (1753) 51, p.p.; Gren. Monogr. Cerast. (1841) 25; Fenzl in Ledeb. Fl. Ross. 1 (1842) 404. —Ic.: Fl. Tadzh. SSR, 3, Plate 79, fig. 6.

Described from around Paris. Type in Paris (P).

Along banks of brooks, in sparse forests.

Note. In all probability, it may be found in west. Kashgar and Junggar.

6. **C. lithospermifolium** Fisch. in Mem. Soc. natur. Moscou, 3 (1812) 80; Ser. in DC. Prodr. 1 (1824) 419; Gren. Monogr. Cerast. (1841) 17; Fenzl in Ledeb. Fl. Ross. 1 (1842) 399; Kar. et Kir. in Bull. Soc. natur. Moscou, 15, 1 (1842) 174; Turcz. in Bull. Soc. natur. Moscou, 15 (1842) 611; Trautv. in Bull. Soc. natur. Moscou, 32, 1 (1860) 161; Kryl. Fl. Zap. Sib. 5 (1931) 1010; Schischk. in Fl. SSSR, 6 (1936) 441; Grub. Konsp. fl. MNR [Conspectus of Flora of People's Republic of Mongolia] (1955) 128; id. Opred. rast. Mong. [Key to Plants of Mongolia] (1982) 101; Ikonnik. Opred. rast Pamira [Key to Plants of Pamir] (1963) 107; Opred. rast. Sr. Azii [Key to Plants of Mid. Asia] 2 (1971) 235; Claves pl. Xinjiang. 2 (1983) 240. —Ic.: Fl. Tadzh. 3, Plate 78, fig. 10; Grub. Opred. rast. Mong. [Key to Plants of Mongolia] Plate 142, fig. 198.

Described from Siberia. Type in St.-Petersburg (Linn.).

In alpine meadows, alpine steppes, rock talus.

IA. Mongolia: *Mong. Alt.* ("Kharkhira; upper Sagliin-gola"—Grub. l.c.).

IB. Kashgar: *Nor.* (Uch-Turfan, June 30, 1908—Divn.).

IIA. Junggar: *Jung. Alat.* (Kazan crossing, Aug. 10, 1878—A. Reg.), *Tien Shan* (in valleys of Yaak-Tash and on Suyuk, Sarz and Barskaun crossings, July 1872—A. Reg.).

General distribution: Nor. Tien Shan; Mid. Asia, West. Sib. (Altay), East. Sib. (Sayans), Nor. Mong. (Fore Hubs., Hang.).

7. **C. maximum** L. Sp. pl. (1753) 439; Ser. in DC. Prodr. 1 (1824) 415; Gren. Monogr. Cerast. (1841) 15; Fenzl in Ledeb. Fl. Ross. 1 (1842) 399; Turcz. in Bull. Soc natur. Moscou, 15, 3 (1842) 612; Kryl. Fl. Zap. Sib. 5 (1931) 1008; Schischk. in Fl. SSSR, 6 (1936) 440; Kamelin et al in Byull. Mosk. obshch. isp. prir. otd. biol. 90, 5 (1985) 115. —Ic.: Fl. SSSR, 6, Plate 22, fig. 11.

Described from Siberia. Type in London (Linn.).

In forests, riverine scrubs, meadows.

IIA. Junggar: *Tien Shan* (south. slope of Tien Shan, near brook, June 5, 1879—Przew.), *Jung. Gobi* (Baga-Bogdo mountain range, Budun-Kharchaityn-Gol basin, on border with China, subalpine meadow, 2000 m, July 28, 1979—Gub.).

General distribution: Arct. (Europ., Asian), West. Sib., East. Sib., Far East, Nor, Amer.

Note. The location of this species in Tien Shan is considerably disjointed from its main range of distribution.

8. **C. pauciflorum** Stev. ex Ser. in DC., Prodr. 1 (1824) 414; Ledeb. Fl. alt. 2 (1830) 176; Gren. Monogr. Cerast. (1841) 19; Schischk. in Fl. SSSR, 6

(1936) 439; Grub. Konsp. fl. MNR [Conspectus of Flora of People's Republic of Mongolia] (1955) 128; Opred. rast. Sr. Azii [Key to Plants of Mid. Asia] 2 (1971) 235; Grub. Opred. rast. Mong. [Key to Plants of Mongolia] (1982) 101; Claves pl. Xinjiang. 2 (1983) 241. —*C. ledebourianum* Ser. in DC. Prodr. 1 (1824) 420. —*C pilosum* Ledeb. in Mem. Ac. Sci. St.-Petersb. 5 (1815) 539, non Sibth. et Sm.; id. Fl. alt. 2 (1830) 178; Gren. Monogr. Cerast. (1841) 19; Fenzl in Ledeb. Fl. Ross. 1 (1842) 398; Turcz. in Bull. Soc. natur. Moscou, 15, 3 (1842) 611; Hemsl. in J. Linn. Soc (London) Bot. 23 (1886) 67; Maxim. Enum. pl. Mong. (1899) 103.

Described from Siberia. Type in Geneva (G).

In forests, scrubs.

IA. Mongolia: *Mong. Alt.* ("Kobdo river"— Grub. l.c.).

IIA. Junggar: *Cis-Alt.* (Kungeity river valley, tributary of Kara-Irtysh, in forest meadow, July 8, 1908—Sap.; 30 km nor. of Koktogai, right bank of Kairta river, valley of Kuidyn river, mixed spruce-larch forest, July 15, 1959—Yun. et al; in Burchum region, in forest, No. 3287, Sept. 6, 1959—Lee and Chu; in Qinhe region, No. 1246, Aug. 2, 1956—Ching).

General distribution: Europe, Mid. Asia, West. Sib. (Altay), East. Sib. (Sayans), Far East, Nor. Mong. (Fore Hubs., Hent., Hang.), China (Nor., Nor.-West.), Japan.

9. **C. pumilum** Curt. Fl. Lond. 2, 6 (1777) vol. 30; Gren. Monogr. Cerast. (1841) 33; Hemsl. in J. Linn. Soc. (London) Bot. 36 (1904) 464; Will. in J. Linn. Soc. (London) Bot. 38 (1907–1909) 397; Sell et Whitehead in Fl. Eur. 1 (1964) 144. —*C. tetrandrum* Curt. l.c., vol. 31.

Described from Europe. Type in Paris (P) ? Plate 5, fig. 2.

In alpine meadows.

IIIA. Qinghai: *Nanshan* (Yamatyn-umru mountains, alpine meadow, June 23, 1894—Rob.).

General distribution: Europe, Himalayas (west.).

Note. C. Grenier (l.c.) pointed out that, among plants of this species with 5-merous flowers, specimens with 4-merous flowers (4 stamens and 4 styles as well) occur. The latter probably represent a specific race.

10. **C. pusillum** Ser. in DC. Prodr. 1 (1824) 418; Kar. et Kir. in Bull. Soc. natur. Moscou, 15, 1 (1842) 174; Kryl. Fl. Zap. Sib. 5 (1931) 1013; Schischk. in Fl. SSSR, 6 (1936) 458; Hao in Bot. Jahrb. 68 (1938) 595; Grub. Konsp. fl. MNR [Conspectus of Flora of People's Republic of Mongolia] (1955) 129; Ikonnik. Opred. rast. Pamira [Key to Plants of Pamir] (1963) 107; Opred. rast. Sr. Azii [Key to Plants of Mid. Asia] 2 (1971) 237. —*C. vulgatum* η *leiopetalum* Fenzl in Ledeb. Fl. Ross. 1 (1842) 410. —**Ic.:** Fl. SSSR, 6, Plate 24, fig. 2.

Described from Siberia (based on Fischer's drawing). Type lost. Neotype in St.-Petersburg (LE).

In alpine meadows, rock screes.

IA. Mongolia: *Mong. Alt.* (Khan-Taishiri-ula, gentle slope at tip, Aug. 16, 1945—Leont'ev; upper Ketsu-Sairin-gol river, alpine meadow, July 26; Adzhi-Bogdo mountain range, Ara-Tszuslan—Ikhe-gol water divide, alpine belt, rubble-rock placers, Aug. 7, 1947—Yun.; Tsastu-Bogdo, in upper Dzuilin-gol, 3000—3400 m, sedge-*Cobresia* swamped waste land, June 24, 1971; Adzhi-Bogdo mountain range, south. macroslope, Ikhe-gol rivulet, 3200–3300 m alt., rocky-stony valley flank, Aug. 22, 1979—Grub., Dariima et al), *Gobi Alt.* (Ikhe-Bogdo mountain range, Narin-Khurimt creek valley midportion, between rocks, June 28; same site, melkozem section among rock screes, June 29; same site, flat crest of mountain range, 3700 m, near cirque fringe in upper Bityuten-ama, compact rubble placer, June 29, 1945—Yun.).

IIA. Junggar: *Tarb.* (Dachen, No. 1660, Aug. 14, 1957—Kuan), *Jung. Alat.* (10 km east of Ven'tsyuan', in Taldy, in forest, No. 2096, Aug. 26, 1957—Kuan), *Tien Shan* (along Danyu river, July 21; from Karagoz village to Danyu, No. 1957, July 18; south of Danyu, No. 410, July 21; 6–7 km south of Danyu, on east. slope, No. 497, July 22; along Danyu river, on slope, No. 2118, July 21; between Daban' and Danyu, on slope, No. 2055, July 19–1957, Kuan; Manas river basin, Ulan-Usu river valley 8–10 km beyond confluence with Dzhartas, on moraine, July 19, 1957—Yun. et al).

IIIB. Tibet: *Weitzan* ("Amne Matchin, auf den Abhangen, No. 1096, Sept. 2, 1930"—Hao, l.c.; Dzhagyn-gol river, No. 148, 1900—Lad.).

General distribution: Jung.-Tarb., Tien Shan, East. Pamir; West. Sib. (Altay), East. Sib. (Sayans).

11. **C. vulgatum** L. Fl. suec. ed. 2 (1755) 158; id. Sp. pl. ed. 2 (1764) 627; Gren. Monogr. Cerast. (1841) 38; Fenzl in Ledeb. Fl. Ross. 1 (1842) 480, cum α *brachypetalum*; Edgew. et Hook. f. in Fl. Brit. India, 1 (1875) 228; Maxim. Fl. Tangut. (1889) 93; id. Enum. pl. Mong. (1889) 104; Walker in Contribs. US Nat. Herb. 28, 4 (1941) 613; Grub. Opred. rast. Mong. [Key to Plants of Mongolia] (1982) 101. —*C. caespitosum* Gilib. Fl. Lithuan. 2 (1781) 159, nom. illegit. Hand.-Mazz. Symb. Sin. 7 (1929) 194; Kryl. Fl. Zap. Sib. 5 (1931) 1012; Schischk., in Fl. SSSR, 6 (1936) 455; Walker, l.c.; Grub. Konsp. fl. MNR [Conspectus of Flora of People's Republic of Mongolia] (1955) 128; Opred. rast. Sr. Azii [Key to Plants of Mid. Asia] 2 (1971) 237; Ma Yu-chuan in Fl. Intramong. 2 (1978) 176, cum var. *glandulosum* Wirtgren; Claves pl. Xinjiang, 2 (1983) 239. —*C. triviale* Link, Enum. 1 (1821) 433; Forbes and Hemsl. in J. Linn. Soc. (London) Bot. 23 (1886) 67. —*C. holosteoides* Fries. Nov. Suec. pl. 4 (1817) 52.; Catal. Nepal, vasc. pl. (1976) 46, pro subsp. *triviale* (Link) Moschl var. *angustifolium* (Franch.) Mizushima. —**Ic.:** Grub. Opred. rast. Mong. [Key to Plants of Mongolia] Plate 42, fig. 199.

Described from Sweden. Type in London (Linn.).

In humid forests and their borders, meadows on slopes and floors of valleys, rock crevices, around roads and villages.

IA. Mongolia: *Mong. Alt.* (Tolbo-Nuru, 3200 m, *Cobresia* meadow, Aug. 5, 1945—Yun.; Adzhi-Bogdo mountain range, in upper Ikhe-gol, waterdivide plateau, 3350–3500 m alt., wet tundra, in rock crevices, Aug. 22, 1979—Grub., Dariima et al), *Gobi Alt.* (Ikhe-Bogdo-ula, east, flank of Narin-Khurimt gorge, about 2900 m, on rocks, July 26, 1948—Grub.), *Alash. Gobi* (mont. Alaschan, July 2, 1873—Przew.).

IIA. Junggar: *Cis-Alt.* (in Qinhe region, No. 1024, Aug. 4, 1956—Ching), *Jung. Alat.* (mountains around Toli, in gorge, No. 1142, Aug. 17, 1957—Kuan), *Tien Shan* (Urten-Muzart crossing, Aug. 2; Aksu valley, May 3, 1877; Burdran, July 5; Kutukshi west of

Kul'dzha, June 6, 1878—Fet.; Sairam lake, July 8–12; *Talki brook, July 22; on Dzhagastai mountain, Aug. 11; *Sairam, Kizemchek, July 1877; around Kazan river, June 22; Bel'bulak crossing, May 20; *on Kokkamyr mountain, July; Kumbel', May 30, 1879—A. Reg.; Mittleres Dschanart Tal, June 14–17, 1903; Sudliches Kiukonik Tal, beim unter Tschon Yailak Pass, June 15; lager am Sudrange des Bogdo-ola, Aug. 26–29, 1908—Merzb.; Urumchi region, upper Tasenku river, Biangou locality, spruce forest belt, Sept. 25, 1929—Pop.; Manas river basin, forest belt, along floor of Koi-su valley, small meadow, July 17, 1957—Yun. et al; Urumchi—Nanshan, No. 0573, July 21, 1956—Ching; in Savan area, Datszymyao, in steppe, No. 1252, July 8; In Danyu region, No. 1383, July 16; in mountains south of Nyutsyuan'tszy, No. 640, July 18; Nyutsyuan'tszy, No. 678, July 19; along Danyu river, on slope, No. 2206, July 23; Chzhaosu area, Aksu region, Koptlak mountains, No. 3496, Aug. 14; 20 km south of Ven'tsyuan', No. 1496, Aug. 14; Sin'yuan', in forest, No. 1128, Aug. 22; in Borgate, south of Yakou, No. 1693, Aug. 30; north of Aksu, in forest, No. 1748, Aug. 31; in Tyan'chi lake region, on water, No. 4320, Sept. 19; Guchen area, in Magolyan, in ditch, No. 4398, Sept. 22; on Barkul' lake, in gorge, No. 4945, Sept. 27, 1957—Kuan; on Talakyus canal in Aksu, No. 8464, Sept. 23, alongside Urumchi-Kucha highway, on slope, No. 6054, July 21, 1958—Lee and Chu).

IIIA. Qinghai: *Nanshan* (along Tetung river and in alpine belt of mountains south of this river, June 13–25, 1872—Przew.), *Amdo* (in upper Huang He, in alpine belt around Khagoma, Aug. 1, 1880; Dzhakhar-Dzhargyn mountain range, on rock screes, alpine belt, June 21, 1884—Przew.).

IIIB. Tibet: *Weltzan* (on rocks along Konchyun-chu river, July 1, 1884—Przew.; Burkhan-Budda mountain range, nor. slope, Nomokhun gorge, May 21, 1900—Lad.).

General distribution: Cosmopolitan.

Note. Variety *tianschanica* (Schischk.) Ju. Kozhevn. in Novit. syst. pl. vasc. 22 (1985) 95 has been asterisked (*). Undoubtedly, other varieties can justifiably be distinguished within C. *vulgatum* but, for the time being, we desist from judging their number, rank of separation and, more difficultly, their nomenclature.

5. **Lepyrodiclis** Fenzl

in Endl. Gen. (1840) 966; id. in Ledeb. Fl. Ross. 1 (1842)

1.   Petals entire or slightly excised at tip; sepals mostly narrow-lanceolate (occasionally, broad-lanceolate). Plants 20–50 (or more) cm tall, with. 0.3–0.6 cm broad leaves ....
..................................1. **L. holosteoides** (C.A. Mey.) Fisch. et C.A. Mey.

+   Petals with 4 large teeth at tip; sepals oblong-oval. Plants 5–15 cm tall, with 0.2–0.3 cm broad leaves............**L. quadridentata** Maxim.

1. **L. holosteoides** (C.A. Mey.) Fisch. et Mey. Enum. pl. nov. 1 (1841) 93; Fenzl in Ledeb. Fl. Ross. 1 (1842) 359; Maxim. Fl. Tangut. 1 (1889) 84; id. Enum. pl. Mong. 1 (1889) 95; Deasy in Tibet and Chin. Turk. (1901) 400; Gorshkova in Fl. SSSR, 6 (1936) 480; Grub. Konsp. fl. MNR [Conspectus of Flora of People's Republic of Mongolia] (1955) 129; Opred. rast. Sr. Azii [Key to Plants of Mid. Asia] 2 (1971) 240; Grub. Opred. rast. Mong. [Key to Plants of Mongolia] (1982) 102; Ikonnik. Opred. rast. Pamira [Key to Plants

of Pamir] (1963) 107 Zhou Li-hou in Fl. Xizang. 1 (1983) 661; Claves pl. Xinjiang. 2 (1983) 246; Fl. desert. Sin. 1 (1985) 448. —*Arenaria holosteoides* Edgew. in Hook. f. Fl. Brit. India, 1 (1875) 241; Forbes and Hemsl. in J. Linn. Soc. (London) Bot. 35 (1902) 137; Schmid. in Feddes Repert. 31 (1933) 42; Walker in Contribs. US Nat. Herb. 28, 4 (1941) 612. —*Lepyrodiclis cerastioides* Kar. et Kir. in Bull. Soc. natur. Moscou, 15, 1 (1842) 170. —Ic.: Fl. desert. Sin. tab. 164, fig. 1–4.

Described from Caucasus. Type in St.-Petersburg (LE).

On river banks, exposed slopes in foothills, garbage, plantations.

IA. Mongolia: *Depr. Lakes* (Shargain-Gobi, Khalyun-gol river, fallow land among barley fields, Aug. 16, 1930—Pob.).

IB. Kashgar: *Nor.* (Uch-Turfan, June 9, 1908. Chegil'—Gumbez, in barley fields, June 19, 1909—Divn.; Talakyus canal in Aksu, No. 8421, Sept. 26; mountain road from Bortu to timber works in Khomot, No. 6977, Aug. 3, 1958—Lee and Chu), *West.* (nor. foothill of Kingtau mountains range, Kosh-Kulak settlement, near irrigation ditch in fields, June 10, 1959—Yun and I.F. Yuan'), *South.* (on nor. foothill of Russkii mountain range, near Sokta village, along irrigation ditches, June 16, 1885—Przew.; nor. slope of Russkii mountain range, Kar-sai village, June 13, 1895—Rob.; "Polu, 2460 m"—Deasy, l.c.).

IIA. Junggar: *Tien Shan* (in. Urumchi region, No. 495, May 28, 1957; Tsitai area, Magolyan village, in steppe, No. 4437, Sept. 22, 1957—Kuan; Borgaty crossing, June 7, 1879—Reg.; Nanshan-kou, in fields along irrigation ditches, June 7, 1877—Pot.; in ploughed fields on Edir' river, Aug. 22; nor. slope of Toshu area, Sept. 1–3, 1895—Rob.; "Hantscheng, May 12, 1928, Tornquist"—Schmid. l.c.).

IIIA. Qinghai: *Nanshan* (Yusun-Khatyma crossing, 2700–3000 m, in forest zone, on garbage, July 24; Mountain range south of Tetung river, in wet sites in valleys and on garbage around houses, as also in ploughed fields, July 25, 1872; in fields around Cheibsen temple, July 20, 1880—Przew.).

IIIB. Tibet: *South.* ("Panggong Tso, July 25, 1927, Bosshard"—Schmid, l.c.; "Pulan'"—Zhou Li-hou, l.c.).

IIIC. Pamir (Kshui-ku, weed vegetation near irrigation ditch, Aug. 9; Tashkurgan, in meadow, July 25, 1913—Knorring; Issyk-su river estuary, 3100 m, July 19; same site, unplanted field, July 3, 1942—Serpukhov; valley around Tashkurgan, grassland along irrigation ditches, June 13, 1959—Yun. and I.F. Yuan').

General distribution: Aralo-Casp.; Balk.-Asia Minor, Fore Asia, Caucasus, Mid. Asia, Nor. Mong. (Hang.), China (Nor.-West.), Himalayas (west.).

Note. *L. holosteioides* var. *stellarioides* (Schrenk) Ju. Kozhevn. (see Novit. syst. pl. vasc. 22 (1985) 97) also occurs in Kashgar and Junggar.

L. quadridentata Maxim. Fl. Tangut. 1 (1889) 84; Hemsl. in J. Linn. Soc. (London) Bot. 36 (1904) 489. —*Arenaria quadridentata* Will. in J. Linn. Soc. Bot. (London), 34 (1898–1900) 436.

Described from China (Gansu province). Type in St.-Petersburg (LE). Plate IV, fig. 3. Map 3.

Along river valleys.

General distribution: China (nor.-west.).

Note. In all probability, this species is found in Qinghai.

6. Holosteum L.

Sp. pl. (1753) 88

H. umbellatum L. l.c.; Fenzl in Ledeb. Fl. Ross. 1 (1842) 373, cum α *oligandrum* et β *pleiandrum*; Trautv. in Bull. Soc. natur. Moscou, 32, 1 (1860) 159; Edgew. et Hook. f. in Fl. Brit. India, 1 (1875) 227; Maxim. Enum. pl. Mong. (1899) 102; Gurke in Pl. Eur. 2 (1897) 136; Kryl. Fl. Zap. Sib. 5 (1931) 1017; Murav'eva in Fl. SSSR, 6 (1936) 467; Schischk; in Fl. Turkm. 3 (1948) 28; Vved. in Fl. Uzb. 2 (1953) 359; Orazova in Ill. opr. rast. Kazakhst. [Illustrated Key to Plants of Kazakhstan] 1 (1969) 332; Opred. rast. Sr. Azii [Key to Plants of Mid. Asia] 2 (1971) 238; Claves pl. Xinjiang. 2 (1983) 244. —*Arenaria glutinosa* M.B. Fl. Taur.-cauc. 1 (1808) 344. —*Holosteum liniflorum* Fisch. et Mey. Ind. sem. Horti Petrop. 3 (1836) 39 (non Stev.). —*H. glutinosum* (M.B.) Fisch. et Mey. l.c.; Murav'eva, l.c.; Schischk. l.c. : 29. —*H. polygamum* C. Koch in Linnaea, 15 (1841) 708; Vved. l.c.; Orazova, l.c.; Opred. rast. Sr. Azii [Key to Plants of Mid. Asia] l.c.

Described from Europe. Type in London (Linn.). Map 4.

On sand, rubble slopes and plains, pebble beds.

IIA. **Junggar:** *Tien Shan* (Kara-Choki, April 18; Kuiyankuz, April 19, 1877; Sarybulak around Kul'dzha, April 22; Almaty valley, April 26—1878; Pilyuchi gorge, April 22; Taldy gorge exit, May 15, 1879—A. Reg.), *Jung. Gobi* (submontane plain north of Bogdo-Ula, 45 km east of Urumchi on road to Fukan, badland, April 26, 1959—Yun. et I.F. Yuan'), *Zaisan* (Mai-Kapchagai mountain, rocky slopes, June 6, 1914—Schischk.), *Dzhark.* (around Suidun, May 8, 1878—A. Reg.).

General distribution: Fore Balkh., Jung.-Tarb., Nor. Tien Shan; Europe, Balk.-Asia Minor, Caucasus, Mid. Asia, Himalayas (west.).

7. Sagina L.

Sp. pl. (1753) 128

1. Glabrous perennial plant. Perianth 5-merous; sepals with winglike outgrowth. Capsule dehiscent on 5 valves. Seeds 0.4-0.5 mm long. ....1. **S. saginoides** (L.) Karst.

+ Annual plant, pubescent with simple hairs (with purple septa) admixed with glandular hairs. Perianth 4-merous; sepals without outgrowths. Capsule dehiscent on 4 valves. Seeds 1-1.3 mm long ....2. **S. karakorensis** (Schmid) Ju. Kozhevn.

1. **S. saginoides** (L.) Karst. Deutsche Fl. (1882) 539; Shteinb. in Fl. SSSR, 6 (1936) 472; Vved. in Fl. Uzb. 2 (1953) 360; Opred. rast. Sr. Azii [Key to Plants of Mid. Asia] 2 (1971) 239; Catal. Nepal. vasc. pl. (1976) 46; Zhou Li-hou in Fl. Xizang. 1 (1983) 659; Claves pl. Xinjiang. 2 (1983) 245. — *Spergula saginoides* L. Sp. pl. (1753) 441; Ser. in DC. Prodr. 1 (1824) 394; Turcz. in Bull. Soc. natur. Moscou, 15, 3 (1842) 593. —*Sagina linnaei* Presl,

Rel. Haenk. 2 (1831) 14; Fenzl in Ledeb. Fl. Ross. 1 (1842) 339; Forbes and Hemsl. in. J. Linn. Soc. (London) Bot. 23 (1886) 70; Kryl. Fl. Zap. Sib. 5 (1931) 1018.

Described from Europe and Siberia. Type in London (Linn.).

On rocky slopes, riverine pebble beds.

IIA. Junggar: *Tien Shan* (Aryslyn, July 10, 1879—A. Reg.).

General distribution: Jung.-Tarb., Tien Shan; Arct., Europe, Mediterr., Balk.-Asia Minor, Caucasus, West. Sib., East. Sib., China (North, Nor.-West.), Himalayas, Korean peninsula, Japan.

Note. Plant with 4-5-merous calyx, 8–10 stamens.

2. **S. karakorensis** (Schmid) Ju. Kozhevn. comb. nov. —*Arenaria karakorensis* Schmid in Feddes Repert. 31 (1933) 42. —*A. saginoides* Maxim. Fl. Tangut. 1 (1889) 89; Hao in Bot. Jahrb. 68 (1938) 595; Zhou Li-hou in Fl. Xizang. 1 (1983) 687. —**Ic.:** Fl. Tangut. tab. 31, fig. 1–17.

Described from west. Tibet. Type in Zurich (Z). Plate 4, fig. 6. Plate 3, fig. 3.

On rocks, meadows, along banks and shoals of rivers, up to 5100 m.

IIIB. Tibet: *Chang Tang* (Przewalsky mountain range, on moist sand, Aug. 1890; Kuen Lun', zurge-gol river valley, on clay with pebble, June 22., 1894—Rob.; a specimen of V. Roborowsky (1889) (No. 413) without indication of site of collection is also available; "Kiam, 5100 m, Aug. 11; Aksai-Chin, about 5000 m, Sept. 5, 1927, Bosshard"— Schmid, l.c.), *Weitzan* (on Bychyu river, 3300 m alt., July 9, 1884—Przew.; "Amne Matchin, zwischen steppenplflanzen, No. 1155, Sept. 3, 1930"—Hao, l.c.; "An'do, Chzhunba"—Zhou Li-hou, l.c.).

General distribution: endemic.

Note. In addition, there is a specimen collected by Dutreuill de Rhins at 5570 m alt. on July 20, 1892, without reference to site of collection (apparently Chang Tang). Maximowicz (l.c.) pointed out that 5-merous flowers are also found sometimes in this species but is apparently a rare phenomenon.

8. Minuartia L.

Sp. pl. (1753) 89; Hiern in J. Bot. (London) 37 (1899) 321. —*Alsine* Gaertn. De Fruct. 2 (1891) 223, non L. —*Arenaria* L. Sp. pl. (1753) 423 p.p.

1. Plant entirely glabrous. Peduncles and pedicels 2–5 cm long, capilliform. ....5. **M. stricta** (Swartz) Hiern.
+ Plant pubescent with glandular hairs. Peduncles and pedicels 1–2 cm long, thickened .... 2.
2. Sepals blunt .... 3.
+ Sepals sharp .... 4.
3. Sepals 5–7 mm long; petals twice longer than calyx. Seeds obtuse-tuberculate. Flowers invariably single. Flowering stems up to 10 cm tall. .... 1. **M. arctica** (Stev.) Aschers. et Graebn.

+ Sepals 3–4 mm long; petals as long as calyx or slightly longer. Seeds glabrous or somewhat rugose. Flowers in inflorescence 2–3 or solitary. Flowering stems up to 5 (rarely up to 7) cm tall ....
.... 2. **M. biflora** (L.) Schinz et Thell.

4. Annual plant. Petals 1/3 or 1/2 of calyx ....
.... 4. **M. regeliana** (Trautv.) Mattf.

+ Perennial plants. Petals as long as or longer than calyx .... 5.

5. Sepals 3.5–5 mm long; petals 1.5–2 times longer than calyx, gradually narrowed toward base, slightly emarginate or serrate at tip. Seeds obtuse-tuberculate or subglabrous. Flowering stems up to 20 (average 10) cm tall .... 3. **M. kryloviana** Schischk.

+ Sepals about 3 mm long; petals as long as or slightly longer than calyx, sharply narrowed toward base, rounded at tip. Seeds sharp-tuberculate. Flowering stems up to 10 (average 5–6) cm tall ....
.... 6. **M. verna** (L.) Hiern.

1. **M. arctica** (Stev.) Aschers. et Graebn. Syn. Mitteleur. Fl. 5, 1 (1918) 772; Kryl. Fl. Zap. Sib. 5 (1931) 1023; Schischk. in Fl. SSSR, 6 (1936) 515; Grub. Konsp. fl. MNR [Conspectus of Flora of People's Republic of Mongolia] (1955) 129; id. Opred. rast. Mong. [Key to Plants of Mongolia] (1982) 102. —*Arenaria arctica* Stev. ex Ser. in DC. Prodr. 1 (1824) 404. —*A. muscorum* Fisch. in DC. l.c. 409. —*Alsine arctica* Fenzl, Verbr. Alsin. (1833) 18; id. in Ledeb. Fl. Ross. 1 (1842) 355; Turcz. in Bull. Soc. natur. Moscou, 15, 3 (1842) 589.

Described from Siberia. Type in Geneva (G).

On rocky slopes and talus, on rocks.

IIA. **Junggar:** *Tien Shan* (on Narat mountain in Kunges, at 3000 m alt., No. 9128, Aug. 7, 1958—Lee and Chu; Danyu, No. 1459, July 17, 1957—Kuan).

General distribution: Jung.-Tarb., Nor. and Cent. Tien Shan; Arct., West. Sib. (Altay), East. Sib., Far East, Nor. Mong. (Hent., Hang., Mong.-Daur.), China (Dunbei), Japan.

Note. Some specimens of this species with both 3- and 4-valved capsules.

2. **M. biflora** (L.) Schinz at Thell. in Bull. Herb. Boiss. 2, ser. 7 (1907) 407; Kryl. Fl. Zap. Sib. 5 (1931) 1027; Schischk. in Fl. SSSR, 6 (1936) 516; Grub. Konsp. fl. MNR [Conspectus of Flora of People's Republic of Mongolia] (1955) 129; Opred. rast. Sr. Azii [Key to Plants of Mid. Asia] 2 (1971) 244; Grub. Opred. rast. Mong. [Key to Plants of Mongolia] (1982) 102; Claves pl. Xinjiang. 2 (1983) 242. —*Stellaria biflora* L. Sp. pl. (1753) 422. —*Alsine biflora* Wahlenb. Fl. Lapp. (1812) 128; Fenzl in Ledeb. Fl. Ross. 1 (1842) 355, cum β *carnosula*; Maxim. Enum. pl. Mong. 1 (1889) 96.

Described from Lapland. Type in London (Linn.).

On rocky slopes and talus, alpine grasslands, rarely in arid forests.

IA. Mongolia: *Khobd.* (Dzusylan, in arid forest, July 25, 1879—Pot.), *Mong. Alt.* (Urmogaity pass, June 27, 1903—Gr.-Grzh.; Khargatiin-daba, July 23; Kharagaitu-daba pass, in upper Indertiin-gol, July 24; 2 km from Bulgan somon westward on road to Kharagaitu-khutul, July 24, 1947—Yun.; Tsastu-Bogdo, in upper Dzuilin-gol, 3000–3400 m alt., June 24; upper Kobdo river, Dayan-Nur, nor. slope of Yamatyn-ul, 2350–2500 m alt., July 9, 1971—Grub., Dariima et al; Nariin-gol river valley, upper course, No. 479, July 15, 1984—Dariima, Kamelin).

IIA. Junggar: *Tien Shan* (Narat pass, 2940 m, June 17, 1877—Przew.; Sairam, July 1877; Kumdaban, 3000 m, May 28; 3000 m, June 3; Naryngol, June 10; Arystyn, 2700–3000 m, July 12; in Kokkamyr mountains, 2700 m, July 28; Tyurgan-tsagan, July 1879—A. Reg.; north of Yakou, No. 1428, Aug. 31, 1957—Kuan).

General distribution: Jung.-Tarb., Tien Shan; Arct., West. Sib., East. Sib., Far East, Nor. Mong. (Fore Hubs., Hent.), Nor. Amer.

3. **M. kryloviana** Schischk. in Anim. Syst. Herb. Univ. Tomsk, No. 1 (1930); Kryl. Fl. Zap. Sib. 5 (1931) 1026; Schischk. in Fl. SSSR, 6 (1936) 501; Opred. rast. Sr. Azii [Key to Plants of Mid. Asia] 2 (1971) 243; Claves pl. Xinjiang. 2 (1983) 242. —Ic.: Fl. Kazakhst. 3, Plate 33, fig. 6.

Described from West. Siberia (Altay). Type in Tomsk (TK). Plate 4, fig. 1. Map 3.

On rocky slopes, rocks, coastal pebble beds.

IIA. Junggar: *Cis-Alt.* (in Koktogoi region, No. 1982, Aug. 17, 1956—Ching), *Jung. Alat.* (in Toli area, on top of mountain, No. 1090, Aug. 6, 1957—Kuan), *Tien Shan* (Sairam, Kyzemchek, July 31; Talkibash, July; Aktash on Dzhagastai mountain, Aug.; in Dzhauku valley, Sept. 7, 1877; Bel'bulak pass, May 20; upper Khargos, June; Bogdo mountain, July 24; Pilyuchi gorge, July 28; Aryslyn, July 10, 1879; near Kul'dzha, May 4 and June 4, 1880—A. Reg.; Urumchi region, upper Tasenku river, Biangou locality, Sept. 25, 1929—Pop.; Manas river basin, midportion of Ulan-Usu river valley at confluence of Koi-su river in it, forest belt, July 17; same site, left bank of Ulan-Usu river at its confluence with Dzhartas, subalp. belt, July 18, 1957—Yun. et al; on Danyu river, No. 2203, July 23; Kelisu, July 16; 25 km south-east of Nyutsyuan'tsza, on bank of Nintszyakhe river, No. 35, July 16, 1957—Kuan).

General distribution: Jung.-Tarb.; West. Sib. (Altay), Mid. Asia.

4. **M. regeliana** (Trautv.) Mattf. in Bot. Jahrb. 57, Beibl. 126 (1921) 29; Kryl. Fl. Zap. Sib. 5 (1931) 1020; Schischk. in Fl. SSSR, 6 (1936) 488; Opred. rast. Sr. Azii [Key to Plants of Mid. Asia] 2 (1971) 242; Claves pl. Xinjiang. 2 (1983) 242. —*Alsine tenuifolia* var. *regeliana* Trautv. in Bull. Soc. Natur. Moscou, 42, (1866) 156.

Described from East. Kazakhstan. Type in St.-Petersburg (LE).

Along rivers and near lakes.

IIA. Junggar: *Cis-Alt.* ("Burchum"—Claves pl. Xinjiang. l.c.).

General distribution: Europe, Caucasus, Mid. Asia, West. Sib.

5. **M. stricta** (Sw.) Hiern in J. Bot. (London) 37 (1899) 320; Kryl. Fl. Zap. Sib. 5 (1931) 1025; Schischk. in Fl. SSSR, 6 (1936) 507; Grub. Konsp. fl. MNR [Conspectus of Flora of People's Republic of Mongolia] (1955) 129; id. Opred. rast. Mong. [Key to Plants of Mongolia] (1982) 102. —*Spergula*

stricta Sw. Vetensk. Acad. Handl. Stockh. 20 (1799) 227. —*Alsine stricta* Wahlb. Fl. Lapp. (1812) 127; Fenzl in Ledeb. Fl. Ross. 1 (1842) 357. Described from Scandinavia. Type in Stockholm (S).

IIA. Junggar: *Tien Shan* (Manas river basin, right flank of Koi-su river valley, forest belt, July 17, 1957—Yun. et al).
General distribution: Arct., Europe, West. Sib., East. Sib., Far East, Nor. Mong. (Fore Hubs.), Nor. Amer.

6. **M. verna** (L.) Hiern in J. Bot. (London) 37 (1899) 320; Kryl. Fl. Zap. Sib. 5 (1931) 1022; Schischk. in Fl. SSSR, 6 (1936) 505; Grub. Konsp. fl. MNR [Conspectus of Flora of People's Republic of Mongolia] (1955) 129; Opred. rast. Sr. Azii [Key to Plants of Mid. Asia] 2 (1971) 243; Grub. Opred. rast. Mong. [Key to Plants of Mongolia] (1982) 102; Claves pl. Xinjiang. 2 (1983) 242. —*Arenaria verna* L. Mant. 1 (1767) 72; Ser. in DC. Prodr. 1 (1824) 405. —*A. villosa* Ledeb. in Mem. Ac. Sci. St.-Petersb. 5 (1815) 534. —*A. paniculata* Bge. in Ledeb. Fl. alt. 2 (1830) 168. —*A. costata* Bge. l.c. 170. —*Alsine verna* Wahlb. Fl. Lapp. (1812) 129; Fenzl in Ledeb. Fl. Ross. 1 (1842) 347, cum α *nivalis* et β *alpestris*; Turcz. in Bull. Soc. natur. Moscou, 15, 3 (1842) 590; Maxim. Enum. pl. Mong. 1 (1889) 95.

Described from Europe (Alps). Type in London (Linn.).
In deciduous forests, on alpine grasslands, on rock screes.

IA. Mongolia: *Khobd.* (Ulan-daban, in forest, June 23; in larch forest on Dzusylan, July 12; in forest along Ulan-Nachina valley, July 15; Kharkira river valley near Tyurgun river estuary, July 22, 1879—Pot.; Turgen' mountain range, Turgen'-gola valley 7 km beyond estuary, on upper forest boundary, July 17, 1971—Grub., Dariima et al), *Mong. Alt.* (Dain-gol lake, June 28, 1903—Gr.-Grzh.; Tsagan-gol river, Prokhodnoi river gorge, June 30, 1905—Sap.; Bulgan somon summer camp, on Kharagaitu-Khutul' road, on lower forest boundary, July 27, 1942—Yun.; upper Buyantu river, Chigirtei-gol, 12 km beyond lake, nor. slope of Chigirtei-ul, 2600–2800 m alt., July 4; Kobdo river, Dayan-nur, nor. slope of Yamatyn-ula, 2350–2500 m alt., July 9, 1971—Grub., Dariima et al; Nariin-gol river valley, upper portion, No. 476, July 15, 1984—Dariima, Kamelin).
IIA. Junggar: *Cis-Alt.* (on Ui-Chilik river, July 20, 1876—Pot.), *Tarb.* (Saur mountain range, Karagaitu river valley, Bain-Tsagan creek valley on right bank, subalpine belt, among rocks, June 23, 1957—Yun. and I.F. Yuan'; north of Dachen town, No. 1580, 1565, Aug. 13, 1957—Kuan), *Jung. Alat.* (in Toli area, on water, No. 1376, Aug. 8, 1957—Kuan), *Tien Shan* (Yuldus, May 13, 1877—Przew.; Sairam lake basin, July 12; Burkhan-tau, June 5; Urten-Muzart, Aug. 4, 1878—Fet.; Taldy river (Irenkhabirga), 2780 m, May 26; Kumbel', 2700–3000 m, May 31, 1879; on Borborogussun river, 2700 m, June 15; Chapchal crossing, June 28; Sairam, Talkibash, July 22, 1877; Bogdo mountain, 3000 m, July 25; on Kokkamyr mountains, 2700 m, July; Aryslyn, July 17; Chubaty (Sairam) crossing, 2400–2700 m, Aug. 2; Mengete, 3000—3300 m, July 4; Baibeshan' crossing, Aug. 3, 1880—A. Reg.; Passe zwischen Kin-su und Kurdai, July 3, 1907; Montes Bogdo-ola et opp. Urumtschi, am sudrande des Bogdo-ola, Aug. 26–29, 1908—Merzb.; Urumchi region, upper Tasenku river, Biangou locality, Sept. 25, 1929—Pop.; Manas river basin, Ulan-usu river valley 8–10 km beyond confluence with Dzhartas, alpine belt, July 19, 1957—Yun. et I.F. Yuan'; between Daban and Danyu, No. 2030, July 19; Savan area, Datszymyao, No. 1918, July; 20 km south of Ven'tsyuan', No.

1549, Aug. 14, 1957—Kuan; Tsitai-Beidashan', on shaded slope, No. 5204, Sept. 28, 1959—Lee and Chu).

General distribution: Jung.-Tarb.; West. Siberia (Altay).

Note. Small specimens of this species are indistinguishable from *Minuartia rubella* (Wahlenb.) Hiern.

9. **Arenaria** L.

Sp. pl. (1753) 423

1. Annual plants.. .... 2.
+ Perennial plants.. .... 3.
2. Perianth 4-merous, styles 2, stamens 2. Leaves lanceolate. ....
.... 13. **A. littledalei** Hemsl.
+ Perianth 5-merous, styles 3, stamens 5. Leaves ovoid, cuspidate ..
....26. **A. serpyllifolia** L.
3. Leaf contour orbicular. Shoots elongated, lodging............................
....**A. rotundifolia** M.B.
+ Leaves of different contour. Shoots erect or plants pulvinoid.... 4.
4. Leaves setaceous or linear-lanceolate. Sepals scarious along margin.. .... 5.
+ Leaves oval or ovoid. Sepals without scarious margin or only some of them narrow-scarious...20.
5. Stamens only 5, 5 more transformed into staminodes. Capsule 1–2-seeded. Lower part of plant branched, with thickened nodes and leaf tufts at ends of shoots. Flowers on long pedicels. Cauline leaves about 1 cm long..............................20. **A. pentandra** Maxim.
+ Stamens 10. Capsule many-seeded. Lower part of plants branched but without thickened nodes. Flowers on comparatively short pedicels. Cauline leaves 1.5–2 cm long..6.
6. Stems up to 40–45 cm tall. Cauline leaves longer than internodes...7.
+ Stems not more than 20 cm tall, or plants pulvinoid. Cauline leaves not longer than internodes.. .... 9.
7. Pedicels and sepals 4–5 mm long, densely glandular. Many remnants of old leaves at stem base forming dense mat..................
....10. **A. juncea** M.B.
+ Pedicels and sepals 2–3.5 mm long, glabrous. Remnants of old leaves absent at stem base.. .... 8.
8. Sepals 3–3.5 mm long, with green prominent midnerve.. ....
.... 3. **A. asiatica** Schischk.
+ Sepals 2–3 mm long, evenly straw-coloured, keeled, but without prominent midnerve..14. **A. longifolia** M.B.
9. Pulvinoid plants. Flowers not projecting above cushion surface...10.
+ Plants forming tufts, with peduncles at least twice longer than radical leaves...16.

10. Flowers sessile, without bracts; petals twice longer than sepals. Leaves dense- and long-ciliate along margin.. .... 22. **A. pulvinata** Edgew.
   + Flowers on short pedicels, with lanceolate bracts; petals 1.5 times longer, equal to or slightly shorter than sepals. Leaves sparse- and short-ciliate or glabrous along margin.. .... 11.
11. Flowers 2–2.5 mm long; sepals oval, blunt. Plants glabrous......12.
   + Flowers over 3 mm long; sepals lanceolate; ciliate along margin like leaves.. 13.
12. Leaves 3–4 mm long; midnerve on sepals obscure. Axillary leaf tufts at tips of generative shoots...................6. **A. densissima** Edgew.
   + Leaves 5–6 mm long. Midnerve on sepals distinct. Axillary leaf tufts absent on generative shoots......21. **A. polytrichoides** Edgew.
13. Flowers about 12 mm long; sepals broad-scarious; petals broad-oval, 1.5 times longer than calyx. ....
   .... .7. **A. edgeworthiana** Majumdar.
   + Flowers 4–7 (8) mm long; sepals narrow-scarious; petals narrow-oval or ovoid, generally as long as calyx. ....14.
14. Pedicels pubescent with long tangled hairs. Leaves in lower part subserrate along margin. Some petals with fine notch at tip.. ....11. **A. kansuensis** Maxim.
   + Pedicels glabrous. Leaves in lower part with sparse fine cilia. All petals orbicular at tip .... 15.
15. Leaves curved, their upper half deflexed from stem at right angle or more, making cushion appear crispate. Pedicels 3–5 mm long .. .. 25. **A. roborowskii** Maxim.
   + Leaves straight, with erect bristles. Pedicels 1–2 mm long ....**A. ischnophylla** Will.
16. Sepals narrow-lanceolate, less than 1.5 mm broad. .... 1. **A. acicularis** Will.
   + Sepals broad, ovoid, 1.5–2.5 mm broad. ....17.
17. Plant 3–5 cm tall. Cauline leaves flat, 2–3 mm broad. Sepals blunt, pubescent along margin with multicellular hairs with purple septa, glabrous on back. .... .22. **A. przewalskii** Maxim.
   + Plants up to 20 cm tall. Cauline leaves setaceous, 1–1.5 mm broad. Sepals cuspidate, pubescent with glandular hairs or glabrous........ .... 18.
18. Radical leaves 2.5–7 cm long. Sepals 4–7 mm long, 2.5–3.5 mm broad. Cauline leaves generally as long as internodes...................... ....5. **A. capillaris** Poir.
   + Radical leaves 0.5–2.5 (3) cm long. Sepals (3) 4–5 (6) mm long, 1.5–2 mm broad. Cauline leaves 1/2 or more shorter than internodes...19.

19. Axils of cauline leaves with leaf tufts. Underground parts of shoots usually elongated, forming plaited tangle with rhizome and surface part of plant. Radical leaves 5–12 (15) mm long. ....
....9. **A. griffithii** Boiss.

+ Axillary tufts of leaves absent. Underground parts of shoots significantly contracted, as though forming caudex. Radical leaves 10–35 mm long. ....18. **A. meyeri** Fenzl.

20. Leaves about 4 (5) mm long...21.

+ Leaves more than 5 mm long .... 22.

21. Plant forming more or less compact cushions. Flowers sessile, 4–5-merous, with petals shorter than calyx. Capsule dehiscent on 3–6 valves. Stems without wings. All sepals acute, glabrous like leaves. ....4. **A. caespitosa** (Camb.) Ju. Kozhevn.

+ Plant forming small (less than 3 cm in diam.) tufts. Flowers on pedicels pubescent with eglandular and glandular hairs, only 5-merous, with petals as long as calyx or slightly longer. Capsule dehiscent on 5–6 valves (sometimes not along sutures). Stems with wings. Sepals acute and blunt, ciliate like leaves, and glandular-pubescent along margin, rarely on back..
....12. **A. ladyginii** Ju. Kozhevn.

22. Flowers large, with 7–13 mm long petals, on peduncles up to 5 mm long. Shoots dense-foliate, specially in upper part.. .... 23.

+ Flowers not large, with petals less than 7 (8) mm long, on peduncles longer than 1 cm. Leaves mainly radical.. .... 24.

23. Styles 2. Sepals pubescent with eglandular and glandular hairs. Petals narrow, orbicular at tip. Filaments sometimes ciliate. ....
....6. **A. cerastiiformis** Will.

+ Styles 3. Sepals glabrous or with solitary cilia. Petals broad-oval, often with notch at tip. Filaments invariably glabrous. ....
....24. **A. ramellata** Will.

24. Plant subglabrous (internodes and lower parts of leaves only sometimes ciliate). Tufts 1.5–3 cm tall. Peduncles not longer than 1 cm. ....27. **A. stracheyi** Edgew.

+ Plants distinctly pubescent. Stems longer than 3 cm. Peduncles longer than 1 cm..25.

25. Leaves lanceolate, cuspidate. Pubescence consists of eglandular hairs with colourless septa, all around on peduncles, on both sides of stem on internodes. Anthers light-coloured. ....
.... .18. **A. monantha** Will.

+ Leaves ovoid, broad- or narrow-oval, cuspidate or blunt. Pubescence consists of glandular hairs with wine-coloured septa. Anthers wine-coloured. ....26.

26.   Leaves narrow-oval, pubescent above and beneath. Stem and pedicels with dense glandular pubescence. ....
....2. **A. amdoensis** L.H. Zhou.

+   Leaves broad-oval or ovoid. Stem and pedicels with sparse glandular pubescence or glabrous.. ..27.

27.   Styles 3. Hairs with coloured septa few, mainly on leaves. Leaves ovoid, cuspidate, with sparse, large, rigidulous hairs on both sides or only along margin. Sepals green, cuspidate, cariniform, rather often, narrow-scarious on margin. Seeds not in scarious sac. .....
....8. **A. glanduligera** Edgew.

+   Styles 2 (rarely 3). Hairs with coloured septa prevailing, mainly on pedicels and sepals. Leaves oblong-oval, blunt, with soft cilia along margin. Sepals dark, flat, without scarious margin. Seeds in scarious sac. ....28.

28.   Sepals broad, blunt. Petals 1.5 times (rarely twice) longer than calyx. Pedicels extended above foliated part of stems for 1/3–1/2 of length. ....17. **A. melandryoides** Edgew.

+   Sepals lanceolate, usually rounded at tip. Petals twice longer than calyx. Pedicels many times shorter than foliated part of stem. ....
....29.

29.   Oblong-oval radical leaves no different from cauline leaves. Petals emarginate at tip. Fusiform roots present. ....
....15. **A. melanandra** (Maxim.) Mattf.

+   Spatulate-linear radical leaves in tufts differ from lanceolate cauline leaves. Petals rounded at tip. Fusiform roots absent........
....16. **A. melandryiformis** Will.

1. **A. acicularis** Will. in J. Linn. Soc. (London) Bot. 38 (1909) 400; Majumdar in Bull. Bot. Surv. India, 15, 1–2 (1973) 41.
Described from Himalayas. Type in London (K). Map 5.
On rubbly slopes of mountains.

IIIB Tibet: *South.* (Lhasa, 3600 m, Aug. 28 and Sept. 1904—Waddell; "Phembu Pass, 10-15 miles north of Lhasa, and Yam. Doh Cho, 1904, No. 1118, Walton"— Williams, l.c.; Gyangtse, July-Sept. 1904, No. 95—Walton).
General distribution: Himalayas.

2. **A. amdoensis** L.H. Zhou in Fl. Xizang. (1983) 688.
Described from Qinghai. Type in Hong Kong (HQ).
On mountain slopes up to 4800 m.

IIIA. Qinghai: *Amdo* ("Amdo, alt. 4830 m, Quighai"—Zhou, l.c.).
IIIB. Tibet: *Weitzan* ("An'do"—Zhou, l.c.).
General distribution: endemic.

58

Note. Although the author of this species compares it with *A. saginoides* Maxim., we have placed it in genus *Sagina*. Since we have not studied any specimen of *A. amdoensis* and the description of the species is very brief, its affinity to genus *Sagina* also is not ruled out.

3. **A. asiatica** Schischk. in Anim. syst. herb. Univ. Tomsk, No. 5-6 (1930) 3; Schischk. and Knorr. in Fl. SSSR, 6 (1936) 527; Grub. Konsp. fl. MNR [Conspectus of Flora of People's Republic of Mongolia] (1955) 129; id. Opred. rast. Mong. [Key to Plants of Mongolia] (1982) 102.

Described from West. Siberia. Type in St.-Petersburg (LE).

On sand margins, arid mountain-steppe slopes.

IA. **Mongolia:** *Mong. Alt.* (Dain-gol lake, arid slopes, July 12, 1908—Sap.).
General distribution: West. Siberia (Altay).

4. **A. caespitosa** (Camb.) Ju. Kozhevn. comb. nov. —*Periandra caespitosa* Camb. in Jaguem. Voy. Ind. (1836-39) 27. —*Briomorpha rupifraga* Kar. et Kir. in Bull. Soc. natur. Moscou, 15 (1842) 172. —*Arenaria rupifraga* Fenzl in Ledeb. Fl. Ross. 1 (1842) 780. —*Thylacospermum rupifragum* Schrenk in Fisch. et Mey. Enum. pl. nov. 2 (1842) 53; Trautv. in Bull. Soc. natur. Moscou, 32, 1 (1860) 158; Edgew. et Hook. f. in Fl. Brit. India, 1 (1874) 243; Maxim. Fl. Tangut. (1889) 89; Hemsl. in J. Linn. Soc. (London) Bot. 35 (1902) 170; Will. in J. Linn. Soc. (London) Bot. 38 (1907-1909) 403; Pamp. Fl. Caracor. (1930) 108: Schmid in Feddes Repert. 31 (1933), 42. —*Thylacospermum caespitosum* (Camb.) Schischk. in Sched. ad Herb. Fl. Ross, 9 (1932) 90; Murav'eva in Fl. SSSR, 6 (1936) 545; Opred. rast. Sr. Azii [Key to Plants of Mid. Asia] 2 (1971) 248; Zhou Li-hou in Fl. Xizang. 1 (1983) 704; Claves pl. Xinjiang. 2 (1983) 244. —**Ic.:** Fl. SSSR, 6, Plate 31, fig. 8; Fl. Xizang. 1, tab. 224, fig. 7-12.

Described from India. Type in Paris (P). Plate 7, fig. 4.

On rocks, rocky slopes, moraine and talus, in upper mountain belt, up to 5000 m.

IIA. **Junggar:** *Tien Shan* (Kassan pass, nor. flank, Aug. 12, 1878; on glacier at Taldy pass, May 20, 1879—A. Reg.; Boro-Khoro mountain range, Chichkan, July 12-13, 1899—Korol'kov; Danu-daban crossing, Manas river basin between Ulan-Usu valley and Danu-gol, cushions on rubble screes around pass, July 19; same site, upper Danu-gol river near Se-daban pass, rubble screes near rocks and edge of glacier, July 21, 1957—Yun. et al).
IIIA. **Qinghai:** *Nanshan* (Kuku-Usu river, 3300-3600 m, July 23, 1879—Przew.; Mon'yuan', on moraine at sources of Ganshig river, tributary of Peishikhe river, 3900-4300 m, Aug. 18, 1958—Dolgushin).
IIIB. **Tibet:** *Chang Tang* ("Kiam, Aug. 15; Sirigh-Jilga, Aug. 31; Aksai-Chin, about 5000 m, Sept. 4, 1927, Bosshard"—Schmid, l.c.; "82°45', 35°, 5130 m, June 27, 1896, Wellby et Malcolm; without locality, Deasy et Pike"—Hemsl. l.c.), *Weitzan* (Burkhan-Budda mountain range, south. slope, on beds of erosion and brooks entering Alak-nor lake, 3810 m, May 30, 1900—Lad.; "An'do"—Zhou Li-hou, l.c.), *South.* ("Ban'ge, Gaitsze, Zhitu, Chzhada, Pulan'—Zhou Li-hou, l.c.).
General distribution: Jung.-Tarb., Nor. and Cent. Tien Shan; Himalayas.

5. **A. capillaris** Poir. Encycl. Meth. 6 (1804) 308; Fenzl in Ledeb. Fl. Ross.
1 (1842) 367, cum var. *glabra* et *glandulosa*; Will. in J. Linn. Soc. (London)
Bot. 33 (1897-98) 414; Maxim. Enum. pl. Mong. (1889) 97; Schischk. and
Knorr. in Fl. SSSR, 6 (1936) 530; Grub. Konsp. fl. MNR [Conspectus of Flora
of People's Republic of Mongolia] (1955) 130; id. Opred. rast. Mong. [Key
to Plants of Mongolia] (1982) 102; Ma Yu-chuan in Fl. Intramong. 2 (1978)
161; Zhou Li-hou in Fl. Xizang. 1 (1983) 670, cum var. *glandulosa* Fenzl. —
A. formosa Fisch. ex Ser. in DC. Prodr. 1 (1824) 402; Grub. l.c. 102. —*A.
capillaris* var. *formosa* (Fisch. ex Ser. in DC) Will. in J. Linn. Soc. (London)
Bot. 33 (1898) 414; Fl. Khang. (1989) 104. —**lc.**: Fl. SSSR, 6, Plate 30, fig. 5;
Fl. Intramong. 2, tab. 86, fig. 6-9; Fl. Xizang. 1, tab. 215, fig. 1-6.

Described from Siberia. Type in Geneva (G).

On rocky and rubbly steppe slopes, alpine meadows and rock screes,
up to 5000 m.

IA. Mongolia: *Khobd.* (Ulan-daban, July 22; Dzusylan, on rocks and along inclines
in upper and lower forest margins, July 14; on nor.-east. descent of Bairimen-daban
pass, July 22; in Kharkhira river valley, on rocks, July 10; along south. source of the
same river, around springs, 2400 m, July 22; same site, on rocky descents up to
permanent snow line, July 24—1879, Pot.; upper Tsagan-gol, rocky crest of Ozernii
glacier, on placers, July 6, 1905—Sap.; Turgen' river valley, 2200 m, Aug. 6, 1973—
Banzragch et al), *Mong. Alt.* (Taishir-Ola, in forest, July 15, 1877—Pot.; on top of
Adzhi-Bogdo, in Dzusylyn gorge, Bor-balgasun, July 2, 1903—Gr.-Grzh.; Tolbo-Kungei-
nuru mountain range, upper mountain belt, Aug. 5, 1945—Yun.; Buyantu and Buluguna
water divide, Akhuntyn-daba on Delyun-Kurdzhurtu road, 3050 m, July 2; upper
Kobdo river, Dayan-nur, Yaman-tau mountain, rocky top of north. offshoot, July 9,
1971—Grub., Dariima et al), *Cent. Khalkha* (Delger-Tsogtu somon, 6 km on road south
of somon camp, rubbly plain, Aug. 12, 1950—Lavr.), *East. Mong.* (Manchuria station,
Beishan' mountain, June 23, 1951—S.H. Lee et al), *Depr. Lakes* (in region of Khara-gol
river, on road to Ulyasutai, Sept. 14, 1879—Pot.), *Gobi Alt.* (on south. side of Ubten-
Kotel pass, Aug. 30, 1886—Pot.; Dundu-Saikhan mountains, July 8, 1909—Czet.;
Ikhe-Bogdo mountain range, Bityuten-ama, alpine belt, Aug. 12, 1927—Simukova;
Dundu-Saikhan mountains, on rubbly mountain slope, Aug. 18; Dzun-Saikhan
mountains, on rocks, Aug. 26; Bayan-Tsagan mountains, Khukhu-daban creek valley,
in upper part of slope, among rubble screes, Aug. 11, 1931—Ik.-Gal.; Ikhe-Bogdo
mountain range, plateau in upper Ikhe-Khurimt creek valley, June 28; same site, upper
belt of mountains, rock screes, June 29; same site, flat crest of mountain range, about
3700 m, in upper Bityuten-ama, June 29; same site, Narin-Khurimt-ama and Ketsu-
ama water divide, June 28, 1945—Yun.; Ikhe-Bogdo mountain range, Narin-Khurimt
gorge, on rocks, 2900 m, July 28; same site, plateau-like crest of mountain range, about
3700 m, July 29, 1948—Grub.; Tostu-nuru mountain range, 2205 m, on nor. slope in
desert steppe belt, July 27, 1972—Rachk., Guricheva), *East. Gobi* (27 km south-west of
Mandal-Gobi, on top of conical hill, July 6, 1970—Lavr. et al; 60 km east-south-east of
Mandal-Gobi ajmaq (administrative territorial unit in Mongolia), fine-hummocky massif
nor.-west of Buyant-ula mountains, on top of conical hillock, June 14; 20 km east of
Delger-Khangai somon, on rocky peak, June 14, 1972—Rachk., Guricheva), *Alash. Gobi*
(mont. Alaschan, in rupibus pratique regionis alpinae, July 18, 1873—Przew..).

IB. Kashgar: *East* ("around Khami, Sept. 12, 1875, Pias."—Maxim. l.c.).

IIA. Junggar: *Tien Shan* (mountain descent toward Turkyul' lake, on rubble, June
29, 1877—Pot.).

IIIA Qinghai: *Nanshan* (in alpine Nanshan, 3300-3600 m, July 16, 1879—Przew.).
IIIB Tibet: *South.* ("Lhasa"—Zhou Li-hou, l.c.).
General distribution: Arct. (Asian), West. Sib. (Altay), East. Sib., Far East, Nor.
Mong. (Hent., Hang., Mong.-Daur.), China (Nor., Nor.-east.).

A. cerastiiformis Will. in J. Linn. Soc. (London) Bot. 38 (1909) 402; Walker
in Contribs. US Nat. Herb. 28, 4 (1941) 612; Majumdar in Bull. Bot. Surv.
India, 15. 1-2 (1973) 41; Zhou Li-hou in Fl. Xizang. 1 (1983) 680.
Described from Himalayas (Fari region). Type in London (K) ?
On arid steep bald slopes.
Found probably in east south. Tibet.

General Distribution: China (south.), Himalayas (east.).

Note. For want of adequate material, we are compelled to recognize this
species. It is quite possible that it is only a large-flowered variety of *A.*
melandryoides.

6. **A. densissima** Wall. ex Edgew. in Hook. f. Fl. Brit. India, 1 (1874) 239;
Will., in J. Linn. Soc. (London) Bot. 33 (1897-98) 406 and 38 (1909) 401;
Pamp. Fl. Caracor. (1930) 105; Majumdar in Bull. Bot. Surv. India, 15, 1-2
(1973) 43; Catal. Nepal. vasc. pl. (1976) 45; Zhou Li-hou in Fl. Xizang.
(1983) 673. —*Dolophragma juniperinum* Fenzl in Ann. Wien. Mus. 1 (1836)
63. —**Ic.:** Fl. Xizang. 1, tab. 217, fig. 12-15.
Described from Himalayas. Type in London (K). Plate 8, fig. 5.
On rocky slopes, alpine meadows, rocks.

IIIB. Tibet: *South.* ("Tibet, 1882, No. 1165, 1882, King collector"—Williams, l.c.;
"Nan'mulin"—Zhou Li-hou, l..c).
General distribution: Himalayas (west.).

7. **A. edgeworthiana** Majumdar in J. Ind. Bot. Soc. 44, 1 (1965) 141, non
Buckl.; Zhou Li-hou in Fl. Xizang. 1 (1983) 677. —*A. monticola* Edgew. in
Hook. f. Fl. Brit. India, 1 (1874) 238; Will. in J. Linn. Soc. (London) Bot. 33
(1897-98) 402 and 28 (1909) 400. —**Ic.:** Fl. Xizang. 1, tab. 218, fig. 6-10.
Described from Himalayas (Fari). Type in St.-Petersburg (LE). Plate 7, fig. 1.
On rocky slopes, alpine meadows.

IIIB. Tibett: *South.* ("Lhasa, Nan'mulin, Dintsze"—Zhou Li-hou, l.c.).
General Distribution: Himalayas (east.).

8. **A. glanduligera** Edgew. in Hook. f. Fl. Brit. India, 1 (1874) 240; Will. in
J. Linn. Soc. (London) Bot. 33 (1897-98) 421 and 38 (1909) 401; Marquand
in J. Linn. Soc. (London) Bot. 48 (1929) 165; Majumdar in Bull. Bot. Surv.
India, 15, 1-2 (1973) 41; Catal. Nepal. vasc. pl. (1976) 45; Zhou Li-hou in Fl.
Xizang. 1 (1983) 683.
Described from Himalayas. Type in London: (K). Plate 7, fig. 6.

On rocky and peat-covered slopes of mountains, alpine meadows, riverine pebble beds, 4700-5500 m alt.

IIIB. **Tibet:** *South.* ("Chzhunba"—Zhou Li-hou, l.c.).
General distribution: Himalayas (west., east.).

Note. The quantum of glandular hairs varies widely in this species, sometimes being almost absent; leaves are covered with multicellular cilia which secrete under certain conditions.

Williams (1897-98) rightly pointed out that petals in this species are not always orbicular at tip as pointed out by Edgeworth (Edgew. l.c.): they can be crispate, quite often with a shallow excision, but orbicular form too is seen. Petals in this species have been reported to be dark-red (Marquand, l.c.).

Williams distinguished small plants with blunt sepals as var. *micrantha* Will. in J. Linn. Soc. 38 (1907-1909) 402. Another variety distinguished by him, var. *cernua* Will., is a small form of *A. melanandra* (see R.p. 63, sl. No. 15).

9. **A. griffithii** Boiss. Diagn. ser. 2, 1 (1854) 89; id. Fl. Or. 1 (1867) 697; Henders. et Hume, Lahore to Jarkand (1873) 312; Edgew. in Hook. f., Fl. Brit. India, 1 (1874) 237; Will. in. J. Linn. Soc. (London) Bot. 33 (1897-98) 404; Schischk. and Knorr. in Fl. SSSR, 6 (1936) 536; Fl. Kirg. 5 (1955) 116; Fl. Kazakhst. 3 (1960) 355; Ikonnik. Opred. rast. Pamira [Key to Plants of Pamir] (1963) 108; Opred. Sr. Azii [Key to Plants of Mid. Asia] 2 (1971) 246; Majumdar in Bull. Bot. Surv. India, 15, 1-2 (1973) 43; Podlech und Angers in Mitt. Bot. Staatssaml. Munchen, 13 (1977) 414. —*Eremogone griffithii* (Boiss.) Ikonnik. in Novit. syst. pl. vasc. 10 (1973) 140. —Ic.: Fl. SSSR, 6, Plate 30, fig. 6.

Described from Fore Asia (Afghanistan). Type in Geneva (G).
On rubbly and rocky slopes of mountains.

IIA. **Junggar:** *Tien Shan* (Khanakhai brook south-west of Kul'dzha, June 15, 1878— A. Reg.).
General distribution: Fore Asia, Mid. Asia.

A. ischnophylla Will. in J. Linn. Soc. (London) Bot. 38 (1909) 400; Majumdar in Bull. Bot. Surv. India, 15, 1-2 (1973) 44; Zhou Li-hou in Fl. Xizang. 1 (1983) 676.

Described from Himalayas. Type in London (K).
On rocky slopes of mountains, alpine meadows, 4500-5000 m alt.

General distribution: Himalayas (west.).

Note. This species has not so far been reported from Tibet within the framework adopted in the present work but its find there is highly probable.

62

Although Williams (l.c.) pointed out that this species, like *A. kansuensis*, has pubescent pedicels, the latter are glabrous in the type material. Sizes of sepals and petals cited by Williams (4 and 5 mm, respectively) are only for the type specimens while, in other specimens, they do not differ in size from sepals and petals of *A. roborowskii* and *A. kansuensis*, i.e. reach 7 mm in length.

10. **A. juncea** M.B. Fl. Taur.-cauc. 3 (1819) 309; Fenzl in Ledeb Fl. Ross. 1 (1842) 366; Turcz. in Bull. Soc. natur. Moscou, 15, 3 (1842) 594; Forbes and Hemsl. in J. Linn. Soc. (London) Bot. 23 (1886) 70; Will. in J. Linn. Soc. (London) Bot. 33 (1897-98) 397; Maxim. Enum. pl. Mong. (1889) 97; Schischk. and Knorr. in Fl. SSSR, 6 (1936) 529; Grub. Konsp. fl. MNR [Conspectus of Flora of Mongolian People's Republic] (1955) 130; id. Opred. rast. Mong. [Key to Plants of Mongolia] (1982) 102; Ma Yu-chuan in Fl. Intramong. 2 (1978) 163. —*A. dahurica* Fisch. ex DC. Prodr. 1 (1824) 402. —*Eremogone juncea* Fenzl, Verbreit. Alsin. (1833) 37. —**Ic.**: Fl. SSSR, 6, Plate 29, fig. 5; Fl. Intramong. 2, tab. 86, fig. 1-5.

Described from East. Siberia. Type in St.-Petersburg (LE).

In steppized, meadowy and steppe slopes of mountains, sand steppes, along river valleys, in meadows.

IA. Mongolia: East. Mong. (on exposed sand descents of Muni-ula mountain range, July 26, 1871—Przew.; Kulun-buir-norsk plain, Buin-gol river, on sandy soil, 1899—Pot. and Soldatov).

General distribution: Nor. Mong. (Mong.-Daur., Fore Hing.), China (Dunbei), Korean peninsula.

11. **A. kansuensis** Maxim. in Bull. Ac. Sci. St.-Petersb. 26 (1880) 428; id. Fl. Tangut. (1889) 86; Forbes and Hemsl. in J. Linn. Soc. (London) Bot. 23 (1886) 70; Will. in J. Linn. Soc. (London) Bot. 33 (1897-98) 402, 34 (1898-1900) 434 and 38 (1909) 400; Hand.-Mazz. Symb. Sin. 7 (1929) 195; Rehder and Kobuski in J. Arn. Arb. 14 (1933) 9; Walker in Contribs. U.S. Nat. Herb. 28, 4 (1941) 613; Zhou Li-hou in Fl. Xizang. 1 (1983) 675. —**Ic.**: Maxim. Fl. Tangut. tab. 14, fig. 13.

Described from Qinghai. Type in St.-Petersburg (LE). Plate 8, fig. 3. Map 3.

On alpine meadows, rocks, stone quarries, 4300-5300 m alt.

IIIA. Qinghai: Nanshan (on top of mountain range south of Tetung river, on rocks, July 13, 1879—Przew., typus; on alpine meadows on mountain range between Nanshan and Donkyr, July 21, 1880—Przew.; "alpine region between Radja and Jupar ranges, No. 14157, Rock"—Rehder and Kobuski, l.c.: Wotila: alpine meadows north of Radja, 4200 m, July 1936—Rock; "Ta-Pan-Shan, on an exposed, moist, alpine summit, 1923, Ching"—Walker, l.c.), *Amdo* (in alpine zone not far from Dzhakhar-Dzhargyn mountains, 3150-3300 m, on rocks, June 24, 1880—Przew.).

IIIB. Tibet: South ("Karo La Pass, 15 miles from Lhasa, at 5000 m, No. 1152, 1878, Dungboo"; same site, 4950 m, No. 1149, July 1904—Walton; "Tszyantszy, Lhasa, Datszy, Naidun"—Zhou Li-hou, l.c.).

General distribution: China (Nor.-West., South-West.).

12. **A. ladyginii** Ju. Kozhevn. in Novit. syst. pl. vasc. 21 (1984) 67.
Described from Tibet. Type in St.-Petersburg (LE). Map 5.
On meadow slopes, in high mountains.

IIIB. Tibet: *Weitzan* (right bank of Golubaya river (Yantszy-tszyan), on nor. descent
of pass, on wet humus, about 4700 m alt., No. 396, June 26, 1900—Lad., typus).
General distribution: endemic.

13. **A. littledalei** Hemsl. in. Kew Bull. (1896) 206; id. in J. Linn. Soc.
(London) Bot. 35 (1902) 170; Zhou Li-hou in Fl. Xizang. 1 (1983) 683. —
Gooringia littledalei Will. in Bull. Herb. Boiss. ser. 2, 7 (1897) 530; id. in J.
Linn. Soc. (London) Bot. 38 (1907–1909) 403.
Described from Tibet. Type in London (K). Plate 3, fig. 5.
Habitat not known.

IIIB. Tibet: *Chang Tang* ("Gooring Valley, at 5030 m. Littledale"—Williams, l.c.).
General distribution: endemic.

14. **A. longifolia** M.B. Fl. Taur.-cauc. 1 (1808) 345; Fenzl in Ledeb. Fl.
Ross. 1 (1842) 362; Trautv. in Bull. Soc. natur. Moscou, 32, 1 (1860) 158;
Will. in J. Linn. Soc. (London) Bot. 33 (1897-98) 399; Kryl. Fl. Zap. Sib. 5
(1931) 1029; Schischk. and Knorr. in Fl. SSSR, 6 (1936) 527; Opred. rast. Sr.
Azii [Key to Plants of Mid. Asia] 2 (1971) 245. —*A. otitoides* Adams ex Ser.
in DC. Prodr. 1 (1824) 102. —Ic.: Fl. SSSR, 6, Plate 29, fig. 6.
Described from East. Europe (Volga). Type in St.-Petersburg (LE).
On meadow steppes, solonetzes.

IIA. Junggar: *Dzhark.* (Konur Kuljdsha, June 20, 1843—Schrenk).
General distribution: Aralo-Casp., Fore Balkh., Jung.-Tarb.; Europe, Caucasus,
West. Sib. (Altay).

15. **A. melanandra** (Maxim.) Mattf. apud. Hand.-Mazz. Symb. Sin. 7
(1929) 202; Rehder and Kobuski in J. Arn. Arb. 14 (1933) 9. —*Cerastium
melanandrum* Maxim. in Bull. Ac. Sci. St.-Petersb. 10 (1880) 580; id. Fl. Tangut.
1 (1889) 92; Forbes and Hemsl. in J. Linn. Soc. (London) Bot. 23 (1886) 66;
Zhou Li-hou in Fl. Xizang. 1 (1983) 680. —*Arenaria glanduligera* var. *cernua*
Will. in J. Linn. Soc. (London) Bot. 38 (1909) 402. —*A. paramelanandra* Hara
in J. Jap. Bot. 52, 7 (1977) 1. —Ic.: Maxim. Fl. Tangut. tab. 15, fig. 1-6.
Described from Nor.-West. China (Gansu province). Type in St.-
Petersburg (LE). Plate 7, fig. 2.
On alpine meadows, among shrubs, on rock screes, riverine pebble beds,
3000-4700 m alt.

IIIA. Qinghai: *Nanshan* (south of Tetung river, on nor. slope of mountain range, on
exposed rock screes, 3000-3600 m alt., July 13, 1827 and July 31, 1880—Przew.;
mountain range between Donkyru and Nanshan, on alpine meadows, 3000 m alt., July
21, 1880—Przew.), *Amdo* (not far from Dzhakhar-Dzhargyn mountain, between shrubs,

3000-3600 m alt., June 24, 1880—Przew.; "alpine region between Radja and Jupar ranges, No. 14157, 14379"—Rehder and Kobuski, l.c.; "An'do"—Zhou Li-hou, l.c.). **General distribution:** China (Nor.-West.), Himalayas (west.).

16. **A. melandryiformis** Will. in J. Linn. Soc. (London) Bot. 38 (1909) 399; Majumdar in Bull. Bot. Surv. India, 15, 1-2 (1973) 43; Zhou Li-hou in Fl. Xizang. 1 (1983) 680.

Described from Himalayas (Chumbi region). Type in London (K)? Map 2. On alpine meadows, rock screes, 4600-4800 m alt.

IIIB. Tibet: *South.* ("Lhasa"—Zhou Li-hou, l.c.). **General distribution:** Himalayas (east.).

Note. Williams (l.c.) differentiated 2 forms: glandulosus and hispid, evidently with eglandular hairs.

17. **A. melandryoides** Edgew. in Hook., f. Fl. Brit. India, 1 (1874) 241; Will. in J. Linn. Soc. (London) Bot. 33 (1897-98) 374 and 38 (1909) 399; Marquand in J. Linn. Soc. (London) Bot. 48 (1929) 166; Majumdar in Bull. Bot. Surv. India, 15, 1-2 (1973) 43; Zhou Li-hou in Fl. Xizang. 1 (1983) 680.

Described from Himalayas (Sikkim). Type in London (K). Plate 7, fig. 3. On alpine meadows, 4200-5000 m alt.

IIIB. Tibet: *South.* ("Lhasa, Nan'mulin"—Zhou Li-hou, l.c.). **General distribution:** Himalayas (east.).

18. **A. meyeri** Fenzl in Ledeb. Fl. Ross. 1 (1842) 368, cum *glandulosa;* Trautv. in Bull. Soc. natur. Moscou, 32, 1 (1860) 158; Kryl. Fl. Zap. Sib. 5 (1931) 1093; Schischk. and Knorr. in Fl. SSSR, 6 (1936) 531; Grub. Konsp. fl. MNR [Conspectus of Flora of Mongolian People's Republic] (1955) 130; Opred. rast. Sr. Azii [Key to Plants of Mid. Asia] 2 (1971) 245. —*A. capillaris* var. *meyeri* (Fenzl) Maxim. Enum. pl. Mong. (1889) 98; Will. in J. Linn. Soc. (London) Bot. 38 (1897-98) 415. —*A. androsacea* Grub. in Not. syst. (Leningrad) 17 (1955) 12; Grub. Konsp. fl. MNR [Conspectus of Flora of People's Republic of Mongolia] (1955) 129; id. Opred. rast. Mong. [Key to Plants of Mongolia] (1982) 102.

Described from West. Sib. (Altay). Type lost. On rubbly and rocky steppe slopes of mountains, on rocks.

IA. Mongolia: *Khobd.* (Achit-nura basin, upper Talain-Tologoi-gol 3 km west of Kholbo-nur lake on road to Tsagan-nur from customhouse, rocks, July 14; Tsagan-Shibetu mountain range, Ulan-daba pass, on Tsagan-nur—Ulangom road, Aug. 18, 1971—Grub., Ulzij. et al), *Mong. Alt.* (Shivergin-gol river valley south of Kobdo town, July 20, 1906—Sap.; Khara-Adzarga mountain range, Sakhir-Sala river valley, Aug. 22; Urtu-gol river valley, on rocks of nor. slope, Aug. 17—1930, Pob.; Khan-Taishiri-ula, rock outliers on steep east. slope, 2600-2700 m, Aug. 10, 1945—Leont'ev; 10 km southeast of Yusun-bulak, midportion of nor. trail of Khan-Taishiri mountain range, July 14; Bus-Khairkhan mountain range, upper part of mountain trail; same site, 2-3 km south of Tamchi lake, in valley, July 17—1947, Yun.; Bayan-Undur mountain range, road 20 km south of Bayan-Undur somon, on east. slope of Khatsabchin-Khara-ula, Aug. 26; Bayan-Tsagan, west. extremity of Bayan-Tsagan-ula 21 km from somon on road to Delger somon, Aug. 28; Tamchi-daba pass, about 2400 m, Sept. 4; granite cone on right bank of Buyantu-gol 2 km nor. of Khobdo, on rocks, Oct. 1—1948, Grub.; 9 km south-west of Tsagan-Olom on road to Yusun-Bulak, Aug. 31; Nyutsugin-gol, tributary of Uenchi-gol, 2100 m, June 26; near Tsetseg somon, sand valleys, June 24, 1973— Golubkova and Tsogt; Buyantu river, upper Delyuna, Khesein-khundei valley, near estuary, July 1; *Tsastu-Bogdo-ula in upper Dzuilin-gol, 2850-3000 m, cushion structures, June 24; Kobdo river basin, Duro-nur lake, Mansar-daba crossing on road to Delyun, 2766 m, June 30; Khan-Taishiri mountain range, south. slope in upper Shine-usu, 2380-2450 m, June 18; Khadzhingiin-nuru south of Tsastu-Bogdo, Seterkhi-khutul', 2844 m, June 25, 1971; Khasagtu-Khairkhan, nor. slope of Tsagan-Irmyk-ula, in upper Khunkerin-ama, 2450-2500 m, Aug. 23, 1972; Baga-Ulan-daba crossing, on Must somon to Uenchi somon road, 2845 m alt., Aug. 13, 1979—Grub. et al.; Shadzgat-nuru mountain range, south. slope, Khoit-Dzhargalan-gol river basin, Bayan-sala gorge, July 27; Uirtiin-Khuren-ula mountains, 20 km nor. of Bugat somon and 5 km from crossing, 2543 m, No. 120, July 5, 1984—Dariima, Kamelin), *East. Mong.* (around Gurbunei-bulak, between Kulusutaevsk and Dolon-nor, 1870—Lom.), *Depr. Lakes* (Khan-Khukhei mountain range, south. slope, in lower belt, July 23, 1945— Yun.; on bank of Kharkira river near its emergence from mountains, Sept. 3, 1931— Bar.), *Gobi Alt.* (offshoot of Ikhe-Bogdo mountain range, Bityuten-ama gorge, Oct. 30, 1926—Kozlova; Ikhe-Bogdo mountain range, nor. slopes in upper belt, Aug. 23, 1926—Tug.; Nomogon-ula mountain range, near mountain peak, May 2, 1941; Ikhe-Bogdo mountain range, south. slope, lower part of Artsatuin-ama creek valley, June 7; crossing between Dzun- and Dundu-Saikhan, lower mountain belt; same site, Dundu-Saikhan trail in midportion; same site, 1 km north of crossing; same site, midportion of Dundu-Saikhan above crossing, upper belt, July 22; Ikhe-Bogdo mountain range, upper part of trail near Narin-Khurimt creek valley, Sept. 5, 1943; Dzun-Saikhan mountain range, nor. portion, south. slope of crest; same site, mid. and lower belt, June 19, 1945—Yun; Nemegetu-nuru mountain range, on main peak, 2747 m, around 2600 m, Aug. 8; Tostu-nuru mountain range, main peak of Sharga-Morite, 2300-2565 m, on peak and along nor.-west. slope, Aug. 15; Nemegetu-nuru mountain range, west. extremity, Khara-obo peak, Aug. 7, 1948—Grub.; 15 km nor.-west of Nonogon, July 11, 1972—Guricheva, Rachk.).

IIA. **Junggar:** *Tien Shan* (south. slope, Khanga river valley, 25 km above Balinte settlement on Karashar-Yuldus road, steppe belt, Aug. 1, 1958—Yun., I.-F. Yuan'; Hami, Aug. 31–Sept. 12, 1875—Piasezky (Piassezki)).

IIIA. **Qinghai:** *Nanshan* (Chan-Sai gorge, on clayey slopes, July 23, 1895—Rob.).
General distribution: West. Sib. (Altay), Nor. Mong. (Hang., Mong.-Daur).

19. **A. monantha** Will. in J. Linn. Soc. (London) Bot. 38 (1909) 401; Majumdar in Bull. Bot. Surv. India, 15, 1-2 (1973) 44.
Described from Tibet. Type in London (K).
Habitat not known.

IIIB. **Tibet:** *South.* (Hills above Lhasa, Aug. 1904—Walton, typus !).
General distribution: endemic.

20. **A. pentandra** Maxim. in Bull. Ac. Sci. St.-Petersb. 10 (1878-1880) 580, non Turcz. nom. nud.; id. Enum. pl. Mong. (1889) 96; Will. in J. Linn. Soc. (London). Bot. 33 (1897-98) 373. —*A. potaninii* Schischk. in Fl. URSS, 6

(1936) 536; Opred. rast. Sr. Azii [Key to Plants of Mid. Asia] 2 (1971) 246.
Ic.: Maxim. Enum. pl. Mong. tab. 66, fig. 12-28; Fl. SSSR, Plate 30, fig. 1.
Described from Sinkiang. Type in St.-Petersburg (LE). Plate 8, fig. 6. Map 5.
On desert-steppe slopes and rocks.

IIA. Junggar: *Zaisan* (on Kichkine-tau mountains, near Zaisan post, on rocks, July 27, 1876—Pot., typus !), *Jung. Gobi* (34 km east of Durbul'dzhin on road to Temirtam, on hummocky area in barren steppe, Aug. 6, 1957—Yun. and I-fen' Yuan').
General distribution: endemic.

Note. All new names used by N.K. Turczaninow in Besser's list (Besser, 1834) are illegitimate; consequently, B.K. Schischkin had no ground to replace Maximowicz's epithet published legitimately.

21. **A. polytrichoides** Edgew. in Hook. f. Fl. Brit. India, 1 (1874) 237; Will. in J. Linn. Soc. (London) Bot. 33 (1897-98) 404, 34 (1898-1900) 436 and 38 (1909) 401; Hemsl. in J. Linn. Soc. (London) Bot. 36 (1904) 457; Majumdar in Bull. Bot. Surv. India, 15, 1-2 (1973) 43; Zhou Li-hou in Fl. Xizang. 1 (1983) 679.
Described from Himalayas (Sikkim). Type in London (K). Plate 8, fig. 1.
On alpine meadows, stone quarries, 4200-5400 m alt.

IIIB. Tibet: *Chang Tang* ("Zhitu"—Zhou Li-hou, l.c.), *South.* (Lhasa, alt. 12,000 ft., Aug. 21, 1908—Waddell; "Chzhada, Pulan', Lhasa"—Zhou Li-hou, l.c.).
General distribution: China (South-West.), Himalayas.

22. **A. przewalskii** Maxim. in Bull. Ac. Sci. St.-Petersb. 10 (1878-1880) 578; id. Fl. Tangut. 1 (1889) 70; Rehder and Kobuski in J. Arn. Arb. 14 (1933) 9; Walker in Contribs. U.S. Nat. Herb. 28, 4 (1941) 613. —Ic.: Maxim. Fl. Tangut. tab. 15, fig. 7-18.
Described from Nor.-West. China (Gansu province). Type in St.-Petersburg (LE). Plate 7, fig. 5.
On alpine meadows.

IIIA. Qinghai: *Nanshan* (pratis alpinis jugi a fl. Tetung, July 13-25, 1872—Przew., typus !), *Amdo* (grasslands between Labrang and Yellow River, No. 14508"—Rehder and Kobuski, l.c.).
IIIB. Tibet: *Weitzan* (nor. descent of Chamudug-la pass, 4700 m, July 26, 1900— Lad.; "Amnyi Machen range, No. 14431"—Rehder and Kobuski, l.c.).
General distribution: China (Nor.-West.).

23. **A. pulvinata** Edgew. in Hook. f. Fl. Brit. India, 1 (1874) 238; Will. in J. Linn. Soc. (London) Bot. 33 (1897-98) 405 and 38 (1909) 401; Zhou Li-hou in Fl. Xizang. 1 (1983) 674. —Ic.: Fl. Xizang. 1, tab. 217, fig. 1-6.
Described from Himalayas (Sikkim). Type in London (K). Plate 8, fig. 4.
On rubbly and rocky slopes, alpine meadows, stone quarries, 4300-5000 m alt.

IIIB. Tibet: *South.* ("Dintsze, Nan'mulin, Pulan'"—Zhou Li-hou, l.c.).
General distribution: Himalayas (east.).

24. **A. ramellata** Will. in J. Linn. Soc. (London) Bot. 38 (1909) 399;
Majumdar in Bull. Bot. Surv. India, 15, 1-2 (1973) 44; Zhou Li-hou in Fl.
Xizang. 1 (1983) 684. —*A. forrestii* Diels in Not. Bot. Gard. Edinb. 5 (1912)
181; Hand.-Mazz., Symb. Sin. 7 (1929) 195; Marquand in J. Linn. Soc.
(London) Bot. 48 (1929) 166; Hao in Bot. Jahrb. 68 (1938) 595; Zhou Li-hou
in Fl. Xizang. 1 (1983) 683.

Described from Tibet. Type in London (K). Map 4.

On alpine meadows, 4000-5000 m alt.

IIIB. Tibet: *Weitzan* ("Amne Matchin, 5000 m, No. 1158, Sept. 3, 1930"—Hao, l.c.;
Yantszy-tszyan river basin, left tributary of By-chyu river, 4200 m, on wet alpine meadow
on mountain descent, July 11, 1900—Lad.), *South.* (Karoo-La, 15 miles from Lhasa,
Aug. 13, 1878—Dungboo, typus !).
General distribution: Himalayas (east.).

25. **A. roborowskii** Maxim. Fl. Tangut. (1889) 87; Hemsl. in J. Linn. Soc.
(London) Bot. 36 (1904) 457; Zhou Li-hou in Fl. Xizang. 1 (1983) 670. —*A.
musciformis* Wall. List E. India pl. (1829) No. 641, nomen; Henderson and
Hume, Lahore to Jarkand (1873) 312; Edgew. et Hook. f. in Fl. Brit. India, 1
(1874) 237, non Triana et Planch.; Maxim. l.c. 86; Hemsley and Pearson in
Peterm. Mitt. 28 (1900) 373; Deasy, Tibet and Chin. Turk. (1901) 400; Hemsl.
in J. Linn. Soc. (London) Bot. 35 (1902) 170; Pamp. Fl. Caracor. (1930) 105;
Hedin, S. Tibet, 6, 3 (1922) 84. —*A. polytrichoides* var. *perlevis* Will. in J.
Linn. Soc. (London) Bot. 33 (1897-98) 405 and 38 (1907-1909) 401. —*A.
perlevis* (Will.) Hand.-Mazz. in Oester. Bot. Zeitschr. 79, 1 (1929) 32; Schmid
in Feddes Repert. 31 (1933) 42; Majumdar in Bull. Bot. Surv. India, 15, 1-2
(1973) 43; Catal. Nepal, vasc. pl. (1976) 46. —*A. festucoides* var. *imbricata*
Edgew. et Hook. f. in Fl. Brit. India, 1 (1875) 237; Hedin, l.c. 83. —**Ic.:** Fl.
Tangut. tab. 29. fig. 8-17.

Described from Tibet. Type lost. Plate 8, fig. 2.

On alpine meadows, rocky slopes, stone quarries, riverine alluvium,
4000-5200 m alt.

IIIB. Tibet: *Chang Tang* (Jarkans, plateau at the foot of Karakorum (north-east of
Karakorum Pass), Aug. 10-11, 1856—Schlagintwelt; Przewalsky mountain range, nor.
foothills, on rocks, in talus, Aug. 20; same site, Aug. 21, 1890—Rob.; "Akssu, 4700 m,
1898"—Deasy, l.c.); "inner Tibet, shore of Naktsong-tso, 4636 m, Sept. 11, 1901"—
Hedin, l.c.; "Aksai-Chin, about 5000 m, Sept. 5, 1927, Bosshard"—Schmid, l.c.); *Weitzan*
(on mountains along midcourse of Yangtze river, June 20, 1834, Rob."—Maxim. l.c.
typus?"; on bank of Razboinich'ei river, 4050 m, Aug. 7; on water divide zone of upper
Huang He and Yangtze rivers, June 24—Przew.; Alyk-nor river basin, along beds and
washout zones of brooks, 3600-3900 m, May 30, 1900—Lad.), *South.* (Khamba Fort,
No. 23, between Kangra Lama Pass and Khamba Fort, at 4650 to 5100 m, No. 1163,
1903, Younghusband; "Gooring Valley, 2895 m, Littledale; South shore of Lake Mangtza
Cho, No. 813, Deasy and Picke"—Williams, l.c.; "Naidun"—Zhou Li-hou, l.c.).
General distribution: Himalayas (west., east.).

A. rotundifolia M.B. Fl. Taur.-cauc. 1 (1808) 314; Fenzl in Ledeb. Fl. Ross. 1 (1842) 369; Boiss. Fl. Or. 1 (1867) 700; Schischk. and Knorr. in Fl. SSSR, 6 (1936) 538; Fl. Tadzh. SSR, 3 (1968) 505; Opred. rast. Sr. Azii [Key to Plants of Mid. Asia] 2 (1971) 247. —*A. orbiculata* Royle ex Hook. f. Fl. Brit. India, 1 (1874) 240; Will. in J. Linn. Soc. (London) Bot. 33 (1897-98) 356 and 38 (1909) 398; Hemsl. in J. Linn. Soc. (London) Bot. 36 (1904) 457; Hand.-Mazz., Symb. Sin. 7 (1929) 194; Catal. Nepal, vasc. pl. (1976) 46; Zhou Li-hou in Fl. Xizang. 1 (1983) 685. —*A. turkestanica* Schischk. in Acta Inst. bot. Ac. Sci. URSS, ser. 1, 3 (1936) 172; Schischk. and Knorr. in Fl. SSSR, 6 (1936) 538; Podlech und Angers im Mitt. Bot. Staatssaml. Munchen, 13 (1977) 414. —**Ic.:** Fl. SSSR, 6, Plate 30, fig. 3; Fl. Xizang. 1, tab. 218, fig. 11-17.

Described from Caucasus. Type in St.-Petersburg (LE).

On alpine meadows.

General distribution: Jung.-Tarb., Tien Shan; Balk.-Asia Minor, Caucasus, Mid. Asia, China (South-West.), Himalayas (west., east.).

Note. No herbarium specimen of this species from Central Asia is available but quite probably occurs in Junggar, Kashgar and Tibet.

26. **A. serpyllifolia** L. Sp. pl. (1753) 423; Fenzl in Ledeb. Fl. Ross. 1 (1842) 168; Trautv. in Bull. Soc. natur. Moscou, 32, 1 (1860) 158; Edgew. et Hook. f. in Fl. Brit. India, 1 (1874) 239; Forbes et Hemsl. in J. Linn. Soc. (London) Bot. 23 (1886) 70 and 35 (1902) 137; Maxim. Enum. pl. Mong. (1889) 98; Will. in J. Linn. Soc. (London) Bot. 33 (1897-98) 365 and 34 (1898-1900) 436; Hand.-Mazz., Symb. Sin. 7 (1929) 195; Kryl. Fl. Zap. Sib. 5 (1931) 1034; Schischk. and Knorr. in Fl. SSSR, 6 (1936) 539; Fl. Tadzh. 3 (1968) 506; Opred. rast. Sr. Azii [Key to Plants of Mid. Asia] 2 (1971) 247; Catal. Nepal, vasc. pl. (1976) 46; Podlech und Angers in Mitt. Bot. Staatssaml. Munchen, 13 (1977) 414; Zhou Li-hou in Fl. Xizang. 1 (1983) 688; Ying-Lan, Hazit in Claves pl. Xinjiang. 2 (1983) 237; Kamelin et al in Byull. Mosk. obshch. isp. prir., otd. biol. 90, 5 (1985) 115. —**Ic:** Fl. SSSR, 6, Plate 30, fig. 2.

Described from Europe. Type in London (Linn.).

On rubbly and rocky slopes, along rivers, in fields and on garbage, up to 5000 m.

IIA. Junggar: Tien Shan (San-tas crossing, Tyube river, on sand around river, June 20, 1893—Rob.; Talki gorge, July 2; on Tekes river, Aug. 13; Dzhagastai, Aug. 9, 1877; around Arganaty, May 28, 1878; Mengeto, July 9, 1879—A. Reg.; 15 km north of Nilki-Ulastai road, No. 3949, Aug. 29; Danyu, No. 1452, July 17; Savan area, Datszymyao, No. 1255, July 8; in Savan area, Shichan, No. 3508, Oct. 1, 1956 —Ching), **Jung. Gobi** (nor. extremity of Maikhan-Ulan mountain, 28 km south-west of Bugat somon, Aug. 1, 1984—Kamelin, Dariima).

IIIB. Tibet: *South.* ("Lhasa"—Zhou Li-hou, l.c.).

IIIC. Pamir (Pas-rabat, in barley field, July 3, 1909—Divn.).

General distribution: Aralo-Casp., Fore Balkh., Jung.-Tarb., Tien Shan; Europe, Balk.-Asia Minor, Fore Asia, Caucasus, Mid. Asia, West. Sib., China (Nor., Nor.-West., Cent., East., South-West., South.), Korea, Japan, Africa.

27. **A. stracheyi** Edgew. in Hook. f. Fl. Brit. India, 1 (1874) 240; Will. in J. Linn. Soc. (London) Bot. 33 (1897-98) 374 and 38 (1909) 398; Deasy, In Tibet and Chin. Turk. (1901) 397; Hemsl. in J. Linn. Soc. (London) Bot. 35 (1902) 170; Marquand in J. Linn. Soc. (London) Bot. 48 (1929) 166; Pamp. Fl. Carac. (1930) 105; Schmid in Feddes Repert. 31 (1933) 42; Zhou Li-hou in Fl. Xizang. 1(1983) 684.

Described from Tibet. Type in London (K).

On wet rocky slopes, coastal meadows, along shoals, 4500-5700 m.

IIIB. Tibet: *Chang Tang* ("NW Tibet, 5700 m, in dampish soil on broken granite, Aug. 12, 1896 —Pike, No. 876; Panggong Tso, July 25; Aksai-Chin, about 5000 m, Sept. 5, 1927, Bosshard"—Schmid, l.c.; "Zhitu"—Zhou Li-hou, l.c.), *Weitzan* ("An'do"— Zhou Li-hou, l.c.), *South.* (Rakas Tal, 4590 m, 1848:, Strachey, Winterbottom—typus !; "Pulan'"—Zhou Li-hou, l.c.).

General distribution: Himalayas (west.).

10. **Moehringia** L.

Sp. pl. (1753) 359

1. Sepals oval, rounded or short-cuspidate at tip, 2-3 mm long. Bracts present.. .... 1. **M. lateriflora** (L.) Fenzl.
+ Sepals lanceolate, long-cuspidate, 3-4 (5) mm long. Bracts absent...2.
2. Sepals ciliate along nerve and on margin. Petals 1/2 or 1/3 shorter than calyx. Leaves oval, lower petiolate, upper sessile. Capsule shorter than calyx .... 2. **M. trinervia** (L.) Clairv.
+ Sepals glabrous. Petals twice as long as calyx. Leaves broad-lanceolate; all leaves sessile. Capsule longer than calyx ....
.... 3. **M. umbrosa** (Bge.) Fenzl.

1. **M. lateriflora** (L.) Fenzl, Verbr. Alsin. (1833) 18 and 38; id. in Ledeb. Fl. Ross. 1 (1842) 371; Kar. et Kir. in Bull. Soc. natur. Moscou, 15, 1 (1842) 596; Rgl. in Bull. Soc. natur. Moscou, 35, 1 (1862) 258; Schischk. and Knorr. in Fl. SSSR, 6 (1936) 541; Grub. Konsp. fl. MNR [Conspectus of Flora of Mongolian People's Republic] (1955) 130; id. Opred. rast. Mong. [Key to Plants of Mongolia] (1982) 102; Fl. Intramong. 2 (1978) 163. —*Arenaria lateriflora* L. Sp. pl. (1753) 423; Forbes et Hemsl. in J. Linn. Soc. (London) Bot. 23 (1886) 70. —Ic.: Grub. Opred. rast. Mong. [Key to Plants of Mongolia] Plate 43, fig. 201; Fl. Intramong. 2, tab. 87.

Described from Siberia. Type in London (Linn.).

In forests and scrubs along catchment areas of rivers and on mountain slopes.

IA. **Mongolia:** *Mong. Alt.* (Urtu-gol river valley, Aug. 19; nor. slopes of Khara-Adzarga mountain range, around Khairkhan-Duru, Aug. 25—1930, Pob.), *Gobi Alt.* (Dzun-Saikhan town, in upper creek valley of Yalo, Aug. 26, 1931—Ik.-Gal.).

IIA. **Junggar:** *Cis-Alt.* (Kurtu river, June 19, 1903—Gr.-Grzh.), *Tien Shan* (in Dzhauku valley, April 1, 1877—A. Reg.; 25 km south-east of Nyutsyuan'tsza town, on bank of Nin'tszyakhe river, July 17, 1957—Kuan).

General distribution: Jung.-Tarb., Nor. Tien Shan; Europe, Arct., West. Sib., East. Sib., Far East, Nor. Mong., China (Nor.), Korean peninsula, Japan.

Note. Sepals in this species have sometimes little pubescence but rarely pubescence is significant. These deviations are, however, of no taxonomic importance.

2. **M. trinervia** (L.) Cvairv. Man. Herb. (1811) 150; Fenzl in Ledeb. Fl. Ross. 1 (1842) 371; Kar. et Kir. in Bull. Soc. natur. Moscou, 15, 3 (1842) 173; Kryl. Fl. Zap. Sib. 5 (1931) 1036; Schischk. and Knorr. in Fl. SSSR, 6 (1936) 540; Opred. rast. Sr. Azii [Key to Plants of Mid. Asia] 2 (1971) 248; Claves pl. Xinjiang. 2 (1983) 243. —*Arenaria trinervia* L. Sp. pl. (1753) 423. —**Ic.:** Fl. SSSR, 6, Plate 31, fig. 4.

Described from Europe. Type in London (Linn.).

In forests and shrubs along streams.

IIA. **Junggar:** *Tien Shan* (Talki river gorge, July 18, 1877—A. Reg.).

General distribution: Aralo-Casp., Jung.-Tarb., Tien Shan; Europe, Balk.-Asia Minor, Caucasus, West. Sib., East. Sib.

3. **M. umbrosa** (Bge.) Fenzl, Verbr. Alsin. (1833) 18 and 38; id. in Ledeb. Fl. Ross. 1 (1842) 372; Kar. et Kir. in Bull. Soc. natur. Moscou, 15, 1 (1842) 173; Kryl. Fl. Zap. Sib. 5 (1931) 1037; Schischk. and Knorr. in Fl. SSSR, 6 (1936) 542; Grub. Konsp. fl. MNR [Conspectus of Flora of People's Republic of Mongolia] (1955) 130; Opred. rast. Sr. Azii [Key to Plants of Mid. Asia] 2 (1971) 248; Grub. Opred. rast. Mong. [Key to Plants of Mongolia] (1982) 102; Claves pl. Xinjiang. 2 (1983) 243. —*Arenaria umbrosa* Bge. in Ledeb. Fl. alt. 2 (1830) 173. —*Moehringia lateriflora* var. *umbrosa* Rgl. in Bull. Soc. natur. Moscou, 35 (1862) 260. —*W. lateriflora* auct. non Fenzl; Maxim. Enum. pl. Mong. (1889) 95. —**Ic.:** Fl. SSSR, 6, Plate 31, fig. 2.

Described from West. Siberia (Altay). Type in St.-Petersburg (LE).

In forests, wet meadows and rocks.

IA. **Mongolia:** *Khobd.* (Ulan-Daba, in forest, June 23, 1879—Pot.), *Mong. Alt.* (upper Kobdo river, Dayan-nur, nor. slope of Yamatyn-ul along lower forest boundary, 2350 m, July 10, 1971—Grub., Ulzij. et al; lower Kobdo lake, Tyurgun river valley, forest, July 2, 1906—Sap.), *East. Mong.* ("on nor. slope of Muin-ula, in birch grove, June 25, 1871, Przew."—Maxim. l.c.).

IIA. **Junggar:** *Cis-Alt.* (30 km nor. of Koktogai, right bank of Kairta river, Kuidyn river valley, mixed spruce-larch forest, July 15, 1959—Yun. and I.-F. Yuan'), *Tarb.* (Saur mountain range, south. slope, valley of Karagaitu river, right bank creek valley of Bain-Tsagai, subalpine meadow, June 23. 1957—Yun. and I.-F. Yuan'), *Tien Shan* (Ili river, June 16; Khanakhai mountains, June 16; between Sumbe and Kazan, June 22, 1878;

Aryslan, July 17, 1879—A. Reg.; mountains close to Santash pass, rocks, June 18, 1893—Rob.; Savan area, Datszymyao, in steppe, No. 1287, July 8; in Danyu region, in forest, No. 1411, July 16—1957, Kuan).

General distribution: Jung.-Tarb., Nor. Tien Shan, Cent. Tien Shan; West. Sib. (Altay).

11. **Spergularia** (Pers.) J. et C. Presl

Fl. Cech. (1819) 94. —*Arenaria* sect. *Spergularia* Pers. Synops. (1805) 504 —*Alsine* L. Sp. pl. (1753) 272

1. Capsule twice as long as calyx. All seeds with broad scarious wing .... 3. **S. maritima** (All.) Chiov.
+ Capsule as along as calyx or somewhat (up to 1/3) longer. Seeds without wings or winged only in lower part of capsule .... 2.
2. Flowers 3-6 mm long. Pedicels generally as long as calyx. Seeds in lower part of capsule winged, not sharp-tuberculate ....
.... **S. marina** (L.) Griseb.
+ Flowers up to 3 mm long. Pedicels 2-6 times longer than flowers. All seeds without wings, sharp-tuberculate .... 3.
3. Plant entirely glandulosus. Petals pink. Stamens 2-3 ....
.... 1. **S. diandra** (Guss.) Heldr. et Sart.
+ Plant entirely glabrous. Petals white. Stamens 10 ....
.... 4. **S. segetalis** (L.) G. Don f.

1. **S. diandra** (Guss.) Heldr. et Sart. in Heldr. Herb. Graec norm. (1855) No. 492, 1124; Kryl. Fl. Zap. Sib. 5 (1931) 1040; Gorshkova in Fl. SSSR, 6 (1936) 557; Opred. rast. Sr. Azii [Key to Plants of Mid. Asia] 2 (1971) 250; Claves pl. Xinjiang. 2 (1983) 237; Fl. desert. Sin. 1 (1985) 447. —*Arenaria diandra* Guss. Fl. Sicul. Prodr. 1 (1827) 515. —*A. salsuginea* Bge. in Ledeb. Fl. alt. 2 (1830) 167. —*Spergularia salsuginea* Fenzl in Ledeb. Fl. Ross. 2 (1842) 166. —**Ic.:** Fl. SSSR, 6, Plate 33, fig. 2; Fl. Kazakhst. 3, Plate 35, fig. 4; Fl. desert. Sin. 1, tab. 163, fig. 5-6.

Described from Europe. Type in Naples.

Along rivers, banks of lakes.

IA. Junggar: *Fore Altay* ("Shara-Sume"—Claves pl. Xinjiang. l.c.).

General distribution: Aralo-Casp., Fore Balkh.; Europe, Mediterr., Balk.-Asia Minor, Caucasus, Mid. Asia, West. Sib.

2. **S. marina** (L.) Griseb. Spic. Fl. Rumel. 1 (1843) 213; Monnier et Ratter in Fl. Eur. 1 (1964) 155; Grub. Opred. rast. Mong. [Key to Plants of Mongolia] (1982) 103. —*S. salina* J. et C. Presl, l.c. 93; Maxim. Fl. Tangut. (1889) 94; id. Enum. pl. Mong. (1889) 105; Kryl. Fl. Zap. Sib. 5 (1931) 1041; Gorshk. in Fl. SSSR, 6 (1936) 561; Grub. Konsp. fl. MNR [Conspectus of Flora of Mongolian People's Republic] (1955) 130; Fl. Kirgiz. 5 (1955) 120; Fl. Kazakhst. 3 (I960) 361; Fl. Tadzh. 3 (1968) 509; Opred. rast. Sr. Azii [Key to Plants of Mid.

Asia] 2 (1971) 249; Ma Tu-chuan In Fl. Intramong. 2 (1978) 159; Claves pl. Xinjiang. 2 (1983) 236; Fl. desert. Sin. 1 (1985) 448. —*Arenaria rubra* var. *marina* L. Sp. pl. (1753) 423, p.p. —*A. salina* Ser. in DC. Prodr. (1824) 401. — *Spergularia media* Boiss. Fl. Or. 1 (1867) 733. —*S. media* var. *heterosperma* Fenzl in Ledeb. Fl. Ross. 2 (1844) 168; Forbes et Hemsl. in J. Linn. Soc. (London) Bot. 23 (1886) 70. —**Ic.:** Grub. Opred. rast. Mong. [Key to Plants of Mongolia] Plate 43, fig. 202; Fl. Intramong. 2, tab. 85, fig. 6-10; Fl. desert. Sin. 1, tab. 163. fig. 1-4.

Described from Europe. Type in Praha (PR).

On saline banks of lakes, wet sand along rivers and brooks, on solonchaks and solonchak-like meadows.

IA. Mongolia: *Mong. Alt.* (nor.-west, foothill of Adzhi-Bogdo mountain range, 60 km nor. of Altay settlement, wet grassland around spring, 1600 m, Sept. 8, 1983— Gub.), *East. Mong.* (Khuna province, in Sinbaer-kuyuchi district around water, June 29, 1951—S.H. Li et al), *Depr. Lakes* (along Dzabkhyn river, bank of Baga-nor lake, July 31; same site, on arid solonchaks of Ol'ge-nor lake, Aug. 29, 1879—Pot.; Ubsa lake, July 28, 1915—Tug.; bank of Shargin-gol river, between Shargin-Tsagan-nor and Dzak-obo lakes, Sept. 8, 1930—Pob.), *Val. Lakes* (Orok-nor, wet lagoon edge at 1260 m, No. 317, 1925, Chaney; Orok-nur lake, solonetzes, Aug. 4, 1926—Tug.; west. border of Orok-nur lake near Tsagan-Deris-urto, Sept. 13, 1943—Yun.), *Gobi-Alt.* (Bayan-Tukhum area, Aug. 5, 1931—Ik.-Gal.; same site, solonchak-like lowland, July-Aug. 1933—Simukova; Bayan-Tukhum lake basin, nor.-east. bank, 1440 m, wet solonchak, Sept. 9, 1979— Grub., Dariima et al), *East. Gobi* (Ongin-gol river floodplain 30 km below Khushu-khida, July 17, 1943—Yun.), *West. Gobi* (Dzakhoi-Dzaram area, Aug. 18, 1943—Yun.), *Alash. Gobi* (nor. Alashan, on bank of tiny Kuku-nor lake, Aug. 21, 1880—Przew.; Edzin-gol river, Chzhargalante area, June 17, 1909—Czet.; valley of Edzin-gol river, Bukhan-khub area, June 2, 1926—Glag.), *Ordos* (along Huang He valley, Aug. 12, 1871—Przew.; on solonetzic bank of Chagan-nor lake, Aug. 24, 1884—Pot.).

IB. Kashgar: *Nor.* (Kashgar oasis, near Upal village, around brook, July 10; between Maral-Bashi and Aksu, near Chadyr-Kul' village, Aug. 6; between Kuchei and Kurlei, near Bugur village, Aug. 20—1929, Pop.; near Shan'nash (=Pichan), 2 km south of Lyan'musin village, No. 6656, June 13; in Yuili region, on Lobulok lake, No. 8565, Aug. 8, 1958—Lee and Chu), *West.* (around Yarkend oasis, June 8, 1889—Rob.), *East.* (Chol-Tag mountain range, near Argai-bulak picket, Aug. 31, 1929—Pop.; Baklimots in Toksun, on terrace, No. 7234, June 10, 1958—Lee and Chu).

IC. Qaidam: *Mont.* (Nor. Qaidam, Kurinka area, May 26, 1895—Rob.).

IIA. Junggar: *Cls-Alt.* (south of Shara-Sume town, No. 2739, Sept. 6, 1956—Ching), *Tien Shan* (Khoiyur-Sumun, south of Kul'dzha, May 27, 1877; Balykchi, April 30; Davati crossing, Sairam, 2330 m, Aug. 19; along Kash river, 1000 m, Sept. 6, 1878—A. Reg.; south. fringe of Khami oasis, Bugas village, 530 m, Aug. 19, 1895—Rob.), *Jung. Gobi* (Guchen, Oct. 11, 1875—Pias; Urumchi town, swamp pools in town, Sept. 21, 1929—Pop.; valley of Manas river, from Paotai state farm to Syaeda, No. 811, June 12; 13 km south-west of Syaeda, No. 76, June 12; Savan town, on waste land, No. 1578, June 26; Savan—Paotai, No. 782, June 10—1957, Kuan; Bulgan river valley 30 km west of Bulgan settlement, 1000 m, Aug. 14, 1982—Gub.).

IIIB. Tibet: *Chang Tang* (Przewalsky mountain range, on wet sand, Aug. 1890— Rob.).

General distribution: Aralo-Casp., Fore Balkh., Jung.-Tarb., Tien Shan; Europe, Mediterr., Balk.-Asia Minor, Fore Asia, Caucasus, Mid. Asia, West. Sib., East. Sib., Far East, Nor. Mong. (Mong.-Daur.), China (Nor., Nor.-West., South-West.), Korean peninsula, Japan, Nor. Amer., South Amer.

Note. Some specimens with 3- as well as 4-valved capsules. Some plants with sepals slightly pubescent along midnerve and others with petals shorter than calyx.

3. **S. maritima** (All.) Chiov. in Ann. Bot. (Roma) 10 (1912) 22. —*S. marginata* (DC.) Kitt. Taschenb. Fl. Deutsch. ed. 2 (1844) 1004; Boiss. Fl. Or. 1 (1857) 733; Gorshk. in Fl. SSSR, 6 (1936) 557. —*S. media* (L.) C. Presl, Fl. Sic. (1826) 161; Monnier et Ratter in Fl. Eur. 1 (1964) 155. —*S. media* β *marginata* Fenzl in Ledeb. Fl. Ross. 2 (1844) 168. —*Arenaria maritima* All. Syn. Stirp. Horti Taurin (1773) 35. —*Arenaria (Spergularia) media* Pers. Syn. 1 (1805) 504, non L. —*A. marginata* DC. Fl. France, 5 (1815) 793. —**Ic.:** Fl. SSSR, 6, Plate 33, fig. 3.

Described from Italy. Type in Torino (TO).

On river banks.

IB. Kashgar: *Nor.* (3 km nor.-west of Tsaokhu in Bugur, alongside river bed with vegetation-covered bottomland (tugai), No. 8688, Sept. 1, 1958—Lee and Chu).

General distribution: Europe, Mediterr., Caucasus, Mid. Asia, Afr.

Note. According to P. Monnier (Candollea, 30, 1975), capsules of some Mediterranean populations of this species are at least 1.5 times longer than calyx and seeds sometimes have very narrow wing or entirely without it.

4. **S. segetalis** (L.) G. Don f. Gen. Syst. 1 (1831-1838) 425. —*Alsine segetalis* L. Sp. pl. 1 (1753) 272; Gorshk. in Fl. SSSR, 6 (1936) 562; Opred. rast. Sr. Azii [Key to Plants of Mid. Asia] 2 (1971) 251; Kamelin et al. in Byull. Mosk. obshch. isp. prir., otd. biol. 90, 5 (1985) 112.

Described from Europe. Type in London (Linn.).

On moist alluviums.

IIA. Junggar: *Jung. Gobi* ("30 km west of Bulgan settlement, No. 5787, Aug. 14, 1982"—Gub.).

General distribution: Aralo-Casp.; Europe, Mid. Asia.

12. **Gymnocarpos** Forsk.

Fl. Aegipt.-Arab. (1775) 65; Benth. et Hook. Gen. pl. 3 (1883) 17

1. **G. przewalskii** Bge. ex Maxim. in Bull. Ac. Sci. St.-Petersb. 26 (1880) 502; Grub. Konsp. fl. MNR [Conspectus of Flora of People's Republic of Mongolia] (1955) 131; id. Opred. rast. Mong. [Key to Plants of Mongolia] (1982) 103; Borbov in Bot. zhurn. 54, 10 (1969) 1579; Ma Yu-chuan in Fl. Intramong. 2 (1978) 159; Claves pl. Xinjiang. 2 (1983) 235; Fl. desert. Sin. 1 (1985) 446. —**Ic.:** Grub. Opred. rast. Mong. [Key to Plants of Mongolia] Plate 44, fig. 207; Fl. Intramong. 2, tab. 85, fig. 1-5; Fl. desert. Sin. tab. 164, fig. 9-11.

Described from Mongolia. Type in St.-Petersburg (LE). Plate 9, fig. 5. Map 4.

On rocky slopes of gorges, sand, rocky barren land, along gorges.

IA. Mongolia: *Alash. Gobi* (Alashan-Urgu road, around Nyudun-khuduk well, May 24, 1909—Davydenko; Bortszon-Gobi area, Khaldzan-ula, bushy barren land, along gorges, June 18, 1949—Yun.; same site, rocky flanks of gorges in upper part of mountain trail and in hummocky area north-east of Khaldzan-ula, Sept. 8, 1950—Lavrenko et al; 30 km north-west of Bayan-Khoto, sandy-pebbly plain between Bain-Nor and Bain-Ula mountains, June 12; same site, southern rim of Bain-Nor mountain, June 12; Bain-Ula mountains, June 13, 1958—Petr.; Khaldzan-ula, 8 km north-east of Gashun-Seveul-khuduk well, July 28, 1970—Grub. Ulzij. et al), *West. Gobi* (Beishan' barren land, south. rim of Chernoi Gobi, July 24; Beishan' barren land, 67 km north of An'si town, black conical hills, July 26, 1958—Petr.), *Ordos* (Huang He river, 1872—Przew., typus !), *Khesi* (foothill region of nor. Nanshan, May 12, 1894—Rob.; nor. slope of mountain range toward Sa-chzhou oasis, on rocky arid steppe, June 24; same site, nor. slope of Mogyt area in Chan-sai gorge, June 23, 1895—Rob.; 80 km south-west of Dunkhuan, on pebbly plain in nor. foothill of Altyntag mountain range, 1650 m, Aug. 8, 1958—Petr.).

IB. Kashgar: *Nor.* (south. slope of Ui-tal river gorge, along river on pebble bed and at base of mountain descents, 1889—Rob.; declivitas australis jugi montium Tianschan, vor Abad, May 30; bei Schaichle und Oi-tatur, anfangs und mitte, June; vor dem eingang zum Dschanart Tal, in der hochsteppe auf Schutthugeln, June 14-17, 1903—Merzb.; Uch-Turfan, May 16, 1908—Divn.; Taushkan-dar'ya river valley 30-35 km beyond Uch-Turfan oasis on arid pebble beds, Sept. 18; Kukukurtuk-Gobi barren land, 10 km south of Ak-yar settlement in south-east. part of Uch-Turfan oasis, nearly barren desert, Sept. 17, 1958—Yun. and I-F. Yuan'; 33 km nor. of Lun'shai town (Bugur), No. 8654, Aug. 30, 1958—Lee and Chu), *West.* (upper Kizil-su river (beyond Kashgar), on red sandstones near Shur-bulak village, July 4, 1929—Pop.), *East* (around Khami, 10 km east in Tsiuzyaotszintszy, sandy-pebbly plain at 930 m alt., Oct. 3, 1959—Petr.).

IIA. Junggar: *Jung. Gobi* (foothill of Baitak-Bogdo-nuru mountain range 6 km from Dzyur-tsagan-nur, *Ephedra*-saltwort barren land, along mountain trail, Sept. 15, 1948—Grub.; Tsigai to pasture in Beidashan', No. 5257, Sept. 29; Beidashan', hammada (rocky desert), No. 2374, Sept. 27—1957, Kuan).

General distribution: endemic.

13. Herniaria L.

Sp. pl. (1753) 218

1   Flowers 5-merous. Leaves 3-5 mm long, glabrous....1. **H. glabra** L.
+   Flowers 4-merous. Leaves 5-10 mm long, pubescent with rigid cilia .... 2. **H. polygama** J. Gay.

1. **H. glabra** L. l.c.; Fenzl in Ledeb. Fl. Ross. 2 (1842) 159; Trautv. in Bull. Soc. natur. Moscou, 37, 2 (1866) 5; Kryl. Fl. Zap. Sib. 5 (1931) 1043; Murav'eva in Fl. SSSR, 6 (1936) 567; Grub. Konsp. fl. Mong. [Conspectus of Flora of People's Republic of Mongolia] (1955) 131; Opred. rast. Sr. Azii [Key to Plants of Mid. Asia] 2 (1971) 252; Grub. Opred. rast. Mong. [Key to Plants of Mongolia] (1982) 103; Fl. Tadzh. 3 (1968) 510. —Ic.: Fl. SSSR, 6, Plate 34, fig. 5; Grub. Opred. rast. Mong. [Key to Plants of Mongolia] Plate 43, fig. 203.

Described from Europe. Type in London (Linn.).
On river banks, wet and marshy meadows.

IA. Mongolia: *Mong. Alt.* (Bulgan somon summer camp in upper Indertiin-gol, marshy meadow in alpine belt, July 24, 1947—Yun.; Bulgan-gol river basin, upper Bayan-gol river and waterdivide, July 23 1984—Dariima, Kamelin; upper Dzhelta river, Sanginin-gola valley, Aug. 13, 1979—Gub.).

IIA. Junggar: *Cis-Alt.* (Korumduk mountain around Kurtu river, June 26, 1903—Gr.-Grzh.; south of Koktogai, in Ukagou region, No. 1735, Aug. 2; Qinhe—Daban'shan'-kou, on arid slope, No. 1560, Aug. 8; Qinhe area, near Tsagankhe river, No. 1516, Aug. 8, 1956—Ching).

General distribution: Aralo-Casp., Fore Balkh., Jung.-Tarb., Tien Shan; Europe, Mediterr., Balk.,-Asia Minor, Fore Asia, Caucasus, Mid. Asia, West. Sib.

Note. Some plants with quite a number of bristles as in *H. polygama.*

2. H. polygama J. Gay in Duchar. Rev. Bot. 2 (1847) 371; Kryl. Fl. Zap. Sib. 5 (1931) 1044; Murav'eva in Fl. SSSR, 6 (1936) 571; Opred. rast. Sr. Azii [Key to Plants of Mid. Asia] 2 (1971) 253.

Described from Europe. Type ?
On river banks.

IIA. Junggar: *Zaisan* (Kaba river near Kaba village, tugai, June 16, 1914—Schischk.).
General distribution: Aralo-Casp.; Europe, West. Sib.

14. **Silene** L.

Sp. pl. (1753) 418; Endlich. Gen. pl. (1836-1840) 973

1. Calyx with 20-30 longitudinal nerves .... 2.
+ Calyx with 10 longitudinal nerves .... 3.
2. Annual plant. Leaves broad-lanceolate. Petals red. Calyx coriaceous, tubular in early stage of flower formation mature seeds inflated, dense-glandular along sharply prominent nerves, with long acute teeth.. 6. **S. conoidea** L.
+ Perennial plant. Leaves oval or ovoid. Petals white or pink. Calyx scarious, inflated from beginning, glabrous, with flat nerves and short blunt teeth .... 26. **S. vulgaris** (Moench) Garcke.
3. Annual or biennial plant.. 4.
+ Perennial plant .... 7.
4. Plant 5-12 cm tall. Pubescence of eglandular crispate hairs only on leaf margin. Seeds flat, rounded, broad-winged, about 2.5 mm in diam .... 16. **S. nana** Kar. et Kir.
+ Plant 25-50 (or more) cm tall. Dense pubescence of eglandular and glandular hairs on all parts. Seeds not flat, without wings, about 1 mm in diam .... 5.
5. Pubescence of only eglandular white hairs. Calyx 9-12 mm long. Petals as long as calyx or only a little longer ....
.... 3. **S. aprica** Turcz. ex Fisch. et Mey.

+   Pubescence of mixed eglandular and glandular hairs, the latter often predominating. Calyx 14-25 mm long. Petals 1.5 times longer than calyx .... 6.

6. Calyx less than 20 mm long, with short, broad-scarious teeth. Petals white with yellowish or greenish tinge. Capsule on carpophore 2-3 mm long .... 25. **S. viscosa** (L.) Pers.

+   Calyx 20-25 mm long, with long subulate teeth. Petals pinkish. Capsule subsessile .... 17. **S. noctiflora** L.

7. Petals entire or slightly excised at tip .... 8.

+   Petals divided for not less than 1/2 of limb 14

8. Flowers many (80-100 or more), aggregated in bunches on inflorescence axis, 2-6 mm long, usually unisexual .... 9.

+   Flowers not many (rarely more than 35), not forming bunches, more than 5 mm long, bisexual .... 10.

9. Inflorescence branches of 1st order short so that inflorescence resembles an interrupted spike. Calyx and pedicels with short pubescence. Radical leaves generally 4-5 mm broad ....
    .... 4. **S. borysthenica** (Grun.) Walter.

+   Flowers on long branches of 1st order in loose panicles. Calyx and peduncles glabrous or scabrous. Capsule 4-9 mm long. Radical leaves generally 6 mm broad ....
    .... 28. **S. wolgensis** (Willd.) Bess. ex Spreng.

10. Leaves mainly radical (cauline leaves 2-3 pairs, excluding bracts). Peduncles glabrous. Reduced vegetative shoots in axils of cauline leaves absent or very few. Calyx in flower campanulate ....
    .... 19. **S. pseudotenuis** Schischk.

+   Leaves mainly cauline. Peduncles pubescent. Axils of cauline leaves with reduced vegetative shoots. Calyx in flower conical ..
    .... 11.

11. Calyx pubescent .... 12.

+   Calyx glabrous. .... 13.

12. Calyx 9-15 mm long, pubescent with eglandular hairs. Capsule 7-8 mm long, on pubescent carpophore 5-7 mm long.....................
    .... 7. **S. gebleriana** Schrenk.

+   Calyx 5-8 mm long, pubescent with glandular as well as eglandular hairs. Capsule 3.5-5 mm long, on glabrous, 1 mm long carpophore or subsessile .... **S. orientalimongolica** Ju. Kozhevn.

13. Petals yellowish-greenish. Leaves lanceolate-linear or linear, 2-6 cm long, 2-6 mm broad. Radical vegetative shoots absent or solitary, and upper part of plant with inflorescence differs vaguely from lower .... 21. **S. sibirica** Pers.

+   Petals white or pinkish. Leaves lanceolate, 1-3 cm long, 1.5-3 mm broad. Radical vegetative shoots abundant and upper part of plant with inflorescence differs sharply from lower...9. **S. holopetala** Bge.

14. Axils of cauline leaves with short vegetative shoots. Nodes thickened .... 15.

\+ Vegetative shoots absent in axils of cauline leaves. Nodes not thickened .... 25.

15. Leaves setaceous, linear or lanceolate, enlarged at base, often spreading subperpendicular to stem .... 16.

\+ Leaves oblong, broad-lanceolate or ovoid, not enlarged at base, recurved invariably at acute angle .... 20.

16. Leaves 0.7-1 cm long .... 22. **S. subcretacea** Will.

\+ Leaves 1.5-8 cm long .... 17.

17. Stems complanate in lower part. Calyx 15-25 cm long. Capsule12-15 mm long. Petal lobes dilated up to tip....l. **S. alexandrae** Keller.

\+ Stems terete all along their length. Calyx 8-18 mm long. Capsule 7-10 mm long. Petal lobes linear .... 18.

18. Entire plant with short scabrous pubescence. Calyx tubular, 14-18 mm long. Anther filaments pilose. Capsule on glabrous, 6 mm long caprophore .... 8. **C. claviformis** Litw.

\+ Some parts of plant glabrous. Calyx clavate, 10-15 mm long. Anther filaments glabrous. Capsule on glabrous or pubescent, 4-5 mm long carpophore .... 19.

19. Leaves usually pubescent above and beneath, arcuately deflexed from stem. Axillary vegetative shoots many. Calyx 12-15 mm long, glabrous or slightly pubescent. Capsule 8-10 mm long, on pubescent carpophore .... 2. **S. altaica** Pers.

\+ Leaves pubescent only along margin, obliquely ascending. Axillary vegetative shoots few. Calyx 10-11 mm long, glabrous .. .. 12. **S. lithophila** Kar. et Kir.

20. Stems and leaves greyish due to dense short pubescence .... 21.

\+ Only some parts of plant with very sparse pubescence .... 24.

21. Pubescence glandular. Shoots in upper part of plant repeatedly branched .... **S. adenocalyx** Will.

\+ Pubescence eglandular. Shoots in upper part not branched .... 22.

22. Calyx 18-23 mm long. Seeds 1.6-2 mm in diam., with deep groove on back .... 14. **S. mongolica** Maxim.

\+ Calyx and seeds small-sized .... 23.

23. Rhizome funiform, grey. Calyx 11-16 mm long, dense-pubescent. Petals with coronal scales .... 20. **S. repens** Patr.

\+ Plant with yellow rachis. Calyx 5-6 mm long, glabrous-like peduncles. Coronal scales absent...... .... 13. **S. maximowicziana** Ju. Kozhevn.

24. Calyx 10-11 mm long. Stem usually simple, branched near centre, pubescent with sparse crispate hairs, more dense on peduncles. Carpophore 2 mm long, barely pubescent. Petal limb with 2 tubercles. .... 23. **S. tatarica** (L.) Pers.

+   Calyx 12-18 mm long. Stems glabrous, usually many, intensely branched from base. Carpophore 5-6 mm long, pubescent. Petal limb without tubercles or other formations .... 18. **S. odoratissima** Bge.

25.   Leaves bright-green, 8-10 mm long, setaceous along margin .... 27. **S. waltoni** Will.

+   Leaves dull green, longer than 10 mm, glabrous or ciliate along margin .... 26.

26.   Capsule subsessile. Inflorescence involuted. Leaves with short, sometimes dense hairs above. Calyx 7-8 mm long .... 5. **S. caespitella** Will.

+   Capsule on carpophore. Inflorescence racemose. Leaves glabrous above. Length of calyx highly variable .... 27.

27.   Hairs with purple septa present in pubescence. Calyx with flat bottom, about 12 mm long .... 11. **S. lhassana** (Will.) Majumdar.

+   Hairs with purple septa absent. Calyx with globose or cuspidate bottom .... 28.

28.   Plant dense-pubescent with fine eglandular hairs. Calyx 25-32 mm long, narrow-conical. Carpophore as long as capsule. Petals white or reddish .... 15. **S. moorcroftiana** Wall. ex Rohrb.

+   Plant with both eglandular and glandular pubescence. Calyx shorter than 25 mm. Carpophore either shorter or longer than capsule....29.

29.   Radical leaves 10-18 mm long, forming rosette. Stems 5-10 cm tall. Calyx 20-24 mm long, clavate. Carpophore longer than capsule. Petals white or pink .... 10. **S. pamirensis** (H. Winkl.) Preobr. ex Schischk.

+   Radical leaves 3-10 cm long, not forming rosettes. Stems 20-40 (60) cm tall. Calyx 6-18 mm long, campanulate. Carpophore 1/2 of capsule. Petals yellowish white or brownish, sometimes with ciliate-margined claw .... 24. **S. tenuis** Willd.

S. adenocalyx Will. in J. Linn. Soc. (London) Bot. 38 (1909) 403; Majumdar in Bull. Bot. Surv. India, 15, 1-2 (1973) 42; Zhou Li-hou is Fl. Xizang. 1 (1983) 736. —**Ic.:** Fl. Xizang. tab. 239.

Described from Tibet. Type in London (K). Isotype in St.-Petersburg (LE). On alpine meadows.

Note. This species is known so far only from West. Himalayas (Khamba Pass, 4800 m, July 1904, Walton) adjoining Tibet. Its found in south. Tibet is highly probable.

1. **S. alexandrae** Keller in Trav. Soc. natur. Kasan. 44, 5 (1912) 71; Kryl. Fl. Zap. Sib. 5 (1931) 1061; Schischk. in Fl. SSSR, 6 (1936) 646; Fl. Kazakhst. 3 (1960) 646; Claves pl. Xinjiang. 2 (1983) 260.

Described from East. Kazakhstan (Zaisan lake). Type in St.-Petersburg (LE). Plate VI, fig. 1.

On rocky steppe slopes and rocks.

IIA. Junggar: *Cis-Alt.* (Barbagai, Burchum, No. 2915, Sept. 2, 1956—Ching), *Jung. Alat.* (Taidzhal mountains (Maili mountain range), on brow of small arid gully, about 1200 m alt., July 19, 1953—Mois.; Dzhair mountain range, mountain-steppe belt, 4-5 km south of Yamaty picket on Chuguchak-Shikho road, in rock crevices on granite conical hills, Aug. 4; south-west. extremity of Maili mountain range, 40-42 km nor.-east of Kzyl-Tuz meteorological station (Junggar gateway) toward Karaganda pass, mountain-steppe belt, Aug. 14, 1957—Yun. and I.-F. Yuan'), *Tien Shan* (Ketmen' mountain range, Sarbushin river valley 1 km nor. of settlement on road to Ili from Kzyl-Kure, Aug. 21, 1957—Yun. and I.-F. Yuan'; in Toli area, Myaoergou village, No. 2434, Aug. 4; from Mulei to Dashitou, No. 4459, Sept. 27, 1957—Kuan).

General distribution: Jung.-Tarb., Nor. Tien Shan.

2. **S. altaica** Pers. Synops. pl. 1 (1805) 497; Ledeb. Fl. Ross. 1 (1842) 315; Trautv. in Bull. Soc. natur. Moscou, 32, 1 (1860) 151; Rohrb. Monogr. Silene (1869) 193; Maxim. Enum. pl. Mong. (1889) 90; Kryl. Fl. Zap. Sib. 5 (1931) 1060; Schischk. in Fl. SSSR, 6 (1936) 646; Grub. Konsp. fl. MNR [Conspectus of Flora of People's Republic of Mongolia] (1955) 131; Opred. rast. Sr. Azii [Key to Plants of Mid. Asia] 2 (1971) 267; Grub. Opred. rast. Mong. [Key to Plants of Mongolia] (1982) 104. —*S. fruticulosa* (Pall.) Schischk. in syn. ex Kryl. Fl. Zap. Sib. 5 (1931) 1060, non Sieb., nec M.B.; Claves pl. Xinjiang. 2 (1983) 251. —*Cucubalus fruticulosus* Pall. Reise 2, Anh. (1773) 739.

Described from Siberia (Altay). Type in St.-Petersburg (LE).

On rocky slopes, rocks, barren steppes.

IA. Mongolia: *Mong. Alt.* (right bank of Bulgan-gol river, Dzun-Khadz-ula low mountains, Aug. 19; Bulgan-gol river basin, midcourse of Bayan-gol river, right bank, July 23, 1984—Dariima, Kam.; 15 km nor. of Bulgan settlement, Ikh-Dzhargalant river valley in Bulgan, 2000 m, Aug. 27, 1983—Gub.), *Gobi. Alt.* (foothills of Gichgeniin-Nuru mountain range, 25 km south-west of Bayan-Under settlement, Aug. 22, 1982; Khurkhu-ula mountains, June 28, 1980—Gub.), *East. Gobi* (prickly pillow structures in foothills of Khutag-ula mountain range, 1200 m, July 10, 1982—Gub.).

IIA. Junggar: *Cis-Alt.* (Syaebai, south of Koktogoi in Ukagou vicinity, in arid steppe, No. 1765, Aug. 11, 1956—Ching), *Tarb.* (on rocks near Kotbukha, Aug. 10, 1876—Pot.), *Tien Shan* (Karagol, No. 1945. July 18; in mine region in Dzhagastai, No. 680, Aug. 7, 1957—Kuan), *Jung. Gobi* (Urumchi, foothills of Bogdo-ola mountain range, Sept. 14, 1929—Pop.; nor. slope of Baga-Khabtak-nuru mountain range, under main peak, 1500-1800 m, Sept. 14; Baitak-Bogdo-nuru mountain range, upper Ulyastu-gol gorge nearly 7 km from opening, Sept. 18, 1948—Grub.; in Ebi-nur lake region, No. 4541, Aug. 19. 1957—Kuan; on right slope of Bodonchiin-gol valley (above 1600 m), June 30, 1973 —Golubkova, Tsogt; Baitag-Bogdo mountain range, Budun-Khargaityn-gol valley near Chinese border, 1900 m, July 29, 1979—Gub.; nor. spur of Takhin-Shara-nuru mountain range, about 2000 m, July 11, 1984—Gub.), *Zaisan* (Mai-Kapchagai mountain, June 6, 1914—Schischk.).

General distribution: Aralo-Casp., Fore Balkh., Jung.-Tarb.; Europe, West. Sib. (Altay), East. Sib. (Sayans).

Note. Some plants from Mongolia have petals almost entire or faintly notched at tip and without floral scales. These plants thus combine characteristics of sections *Holopetalae* Schischk. and *Suffruticosae* Rohrb. suggesting indistinct isolation of first from second section. Specimens of this species do not always reliably differ from *S. lithophila* in calyx size. The variation of this species in leaf breadth, sometimes measuring 8-9 mm, should also be noted.

3. **S. aprica** Turcz. ex Fisch. et Mey. in Ind. sem. Horti Petrop. (1835) 38; id. in Bull. Soc. natur. Moscou, 15, 3 (1842) 580; Ledeb., Fl. Ross. 1 (1842) 317; Forbes and Hemsl. in J. Linn. Soc. (London) Bot. 23 (1886) 64; Maxim., Fl. Tangut. (1889) 82; id. Enum. pl. Mong. (1889) 89. —*Melandryum apricum* (Turcz.) Rohrb., Monogr. Silene (1868) 231; Schischk. in Fl. SSSR, 6 (1936) 713; Grub., Konsp. fl. MNR [Conspectus of Flora of People's Republic of Mongolia] (1955) 131; id. Opred. rast. Mong. [Key to Plants of Mongolia] (1982) 104; Ma Yu-chuan in Fl. Intramong. 2 (1978) 180; Zhou Li-hou in Fl. Xizang. 1 (1983) 734; Claves pl. Xinjiang. 2 (1983) 258; Fl. desert. Sin. 1 (1985) 462; Zhou Li-hou in Acta sci. natur. univ. Intramong. 16, 4 (1985) 586. —*Elisanthe aprica* (Turcz.) Peschkova in Fl. Tsentr. Sib. 1 (1979) 328. — Ic.: Fl. Intramong. 2, tab. 96, fig. 1-6; Fl. desert. Sin. 1, tab. 169, fig. 6-7.

Described from East. Siberia (Fore Baikal). Type in St.-Petersburg (LE).

On rubbly steppe slopes, sand and sand-covered steppes, rocks, flanks of gorges.

IA. Mongolia: Cent. Khalkha (Dzhargalanta river basin, between Botog and Agit mountains, gorge, Sept. 5; same site, on rocks, Sept. 11, 1925—Krasch., Zam.; 40 km north of Altyn-Shire somon, in granitic finely hummocky area, Aug. 6, 1971—Isach. and Rachk.), **East.** (Dariganga, eastward of Ongon-elis sand, Khoshun-khuduk well, Sept. 20; 40 km north of Dzamyn-Ude, Motonge mountains, Aug. 30—1931, Pob.; Kulun-Buir-nor, Abder river, June 25; Khuntu lake, June 22, Ul'gen-ikhe river, July 7— 1899, Pot. and Soldatov; vicinity of Baishintyn-sume, Urgo mountain, Aug. 18, 1927— Zam.; Matad somon, steppe, Aug. 4; same site, 12 km west-north-west of somon, Aug. 15—1949, Yun.; Dariganga, Shiliin-Bogdo-Ula, 1650 m, July 11-12, 1985—Gub.; 75 km south-east of Choibalsan town, July 6, 1985—Odbayar), **Depr. Lakes** (10 km west-south-west of Malchin somon, 1700 m alt., July 20, 1973—Banzragch et al), **Gobi Alt.** (Dundu-Saikhan mountains, on river bed, July 13, 1909, Czet.; foothill at Ikhe-Bogdo town, eroded moraine, Aug. 24, 1926—Tug.; Ikhe-Bogdo, Bityuten-ama, in alpine belt, Aug. 12, 1927—Simukova; Barun-Saikhan mountain range, rocky slope, Sept. 20; Bain-Tsagan mountain range, on mountain slope, Aug. 4; same site, in Khukhu-daban creek valley, Aug. 11; Dzolin mountain range, Aug. 8; same site, Sept. 13; Dundu-Saikhan mountain range, Ulan-khunde creek valley, on rocks, Aug. 20—1931. Ik.-Gal.; Noyan-Bogdo-ula, July 25; south. slope of Ikhe-Bogdo mountain range, Sept. 10; nor. slope of Ikhe-Bogdo mountain range, midportion of Bityutea-ama creek valley, Sept. 12—1943. Yun.; Khuryn-Khana-nuru mountains, rocks, July 25, 1972—Guricheva and Rachk.), **East. Gobi** (Jichi Ola, Outer Mongolia, on rocky summit at 1650 m, No. 419, 1925— Chaney; Del'ger-Khangai somon, Khoir-Ul'tszeitu area—Sharangad, barren steppe, Sept. 14, 1930—Kuznetsov; Ikh-Shankhai-nuru mountains, in cleavages of granite rocks, July 8, 1972—Guricheva and Rachk.), **Alash. Gobi** (in central Alashan, on west. slope of entrance to gorge, July 21, 1873—Przew.), *Ordos* (Huang He river, on light sand, Aug. 7; in valley of Alan-Muren settlement; Aug. 22; Gundzhatagai area, Sept. 6; Taitukhai area, Aug. 29—1884, Pot.).

IIA. **Junggar:** *Tien Shan* (Shuvutin-daba settlement north of Sairam-nur basin on road to Borotal, subalpine steppe, Aug. 18, 1957—Yun. et al; Karagol, July 18, 1957—Kuan; in Kolangou, on shaded slope, No. 5848, June 24; Bartu to Khomote timber plant, under spruce grove, Aug. 4. 1958—Lee and Chu).

General distribution: East. Sib. (south.), Nor. Mong.

Note. Specimens collected by Gubanov from Dariganga have dense pubescence of glandular hairs in upper part.

4. **S. borysthenica** (Grun.) Walters in Feddes Repert. 69 (1964) 47; Chater, Walters in Fl. Eur. 1 (1964) 167. —*S. otites* var. *borysthenica* Grun. in Bull. Soc. natur. Moscou, 11, 2 (1868) 126. —*S. parviflora* (Ehrh.) Pers., Syn. 1 (1805) 497, non Moench; Boiss. Fl. Or. 1 (1857) 607; Kryl. Fl. Zap. Sib. 5 (1931) 1062; Schischk. in Fl. SSSR, 6 (1936) 689; Grub. Konsp. fl. MNR [Conspectus of Flora of Mongolian People's Republic] (1955) 131; Opred. rast. Sr. Azii [Key to Plants of Mid. Asia] 2 (1971) 3273; Grub. Opred. rast. Mong. [Key to Plants of Mongolia] (1982) 103; Claves pl. Xinjiang, 2 (1983) 263. —*S. otites* β *parviflora* Rohrb. Monogr. Silene (1868) 200. —*Otites borysthenica* (Grun.) Klokov in Bot. zhurn. AN UkSSR, 5, 1 (1948) 22.

Described from Europe. Type in Moscow (?).

On rubble slopes, sand.

IA. **Mongolia:** *Depr. Lakes* (Borig-Del' fixed sand south-east of Bayan-nur lake, wormwood steppe in ravine, Aug. 25; same site, wormwood hummocky sand, July 25—1945. Yun.).

IIA. **Junggar:** *Zaisan* (lower course of Belezeka river, tugai, June 18, 1914—Schischk.; 10 km north-west of Khabakhe area, on floodplain, in grass, Sept. 18, 1959—Lee and Chu).

General distribution: Aralo-Casp., Fore Balkh.; Europe (east.), West. Sib. (south), East. Sib. (Ang.-Sayan.), Nor. Mong. (Hang.).

5. **S. caespitella** Will. in J. Linn. Soc. (London) Bot. 38 (1909) 403; Majumdar in Bull. Bot. Surv. India, 15, 1-2 (1973) 30; Zhou Li-hou in Fl. Xizang. 1 (1983) 736. —Ic.: Fl. Xizang. tab. 236.

Described from Himalayas. Type in London (K).

On rubbly mountain slopes, 3600-5000 m alt.

IIIB. **Tibet:** *South.* ("Gyangtse, 1904, Walton"—Will. l.c.).

General distribution: Himalayas (east.).

6. **S. conoidea** L. Sp. pl. (1753) 418; Ledeb. Fl. Ross. 1 (1842) 314; Boiss. Fl. Or. 1 (1857) 580; Rohrb. Monogr. Silene (1868) 92; Henders and Hume, Lahore to Yarkand (1873) 312; Edgew. et Hook. f. in Fl. Brit. India, 1 (1874) 218; Forbes and Hemsl. in J. Linn. Soc. (London) Bot. 23 (1886) 65; Maxim. Fl. Tangut. (1889) 82; id. Enum. pl. Mong. (1889) 88; Will. in J. Linn. Soc.

(London) Bot. 32 (1896) 52; Deasy, in Tibet and Chin. Turk. (1901) 400; Schischk. in Fl. SSSR, 6 (1936) 690; Hao in Bot. Jahrb. 68 (1938) 594; Walker in Contribs. U.S. Nat. Herb. 28, 4 (1941) 613; Grub. Konsp. fl. MNR [Conspectus of Flora of Mongolian People's Republic] (1955) 131; Opred. rast. Sr. Azii [Key to Plants of Mid. Asia] 2 (1971) 274; Grub. Opred. rast. Mong. [Key to Plants of Mongolia] (1982) 103; Podlech und Anders in Mitt. Bot. Staatssaml. Munchen, 13 (1977) 416; Zhou Li-hou in Fl. Xizang. 1 (1983) 732; Claves pl. Xinjiang. 2 (1983) 260; Fl. desert. Sin. 1 (1985) 457. — *Cucubalus conoideus* Lam. Fl. France, 3 (1778) 27. —*S. dioica* Fisch. et Mey., Ind. Sem. Horti Petrop. 7 (1830) 60. —*Pleconax conoidea* (L.) Sourkova in Osterr. Bot Z. 119 (1971) 579; Ikonnik. Opred. vyssh. rast. Badakhsh. [Key to Higher Plants of Badakhsh.] (1979) 151. —**Ic.**: Fl. SSSR, 6, Plate 35, fig. 2; Fl. desert. Sin. 1, tab. 166, fig 5.

Described from Europe. Type in London (Linn.).

On rocky steppized slopes, in plantations and along field borders.

IA. **Mongolia:** *Mong. Alt.* ("south-east.—Tsakhir-bulak"—Grub. l.c. 1982).

IB. **Kashgar:** *West.* (Kshui-ku, near irrigation ditch; between Kshui-ku and Yak-kshu, 10 versts—1 verst = 1.067 km—from Yak-kshu, Aug. 9, 1913—Knor.; upper Kizil-su river, beyond Kashgar, vicinity of Egin mountains, Kara-Tasha floodplain, Aug. 2; Sarykol'sk mountain range, Bostan-Terek locality, Aug. 10, 1929—Pop.), *South.* (Russky mountain range, Achan village, 2280 m, in ploughed fields, June 22; in ploughed fields along Nura river, July 23; on Keriya river, June 30—1885, Przew.; slope of Russky mountain range, Kara-sai village, in plantations, June 6, 1890—Rob.; Tokhta-Khon, July 24, 1890—Grombch.).

IC. **Qaidam:** *South.* (not far from Burkhan-Budda base, in ploughed fields, Aug. 29, 1884—Przew.; nor. slope of Burkhan-Budda mountain range, Khatu gorge, in cultivated and abandoned fields, 3150 m, July 3, 1901—Lad.).

IIA. **Junggar:** *Jung. Gobi* (around Adak village, in fields, June 17, 1877—Pot.; around Urumchi town, resort 5-6 km from town, April 27, 1952—Mois.).

IIIA. **Qinghai:** *Nanshan* (between Choibsen temple and South-Tetung mountain range, 2100-2500 m, in ploughed fields, July 27. 1880—Przew.).

IIIC. **Pamir** (Ulugtuz gorge in Charlym river basin, very many in barley plantations, June 23, 1909—Divn.).

General distribution: East. Pam.; Mediterr., Balk.-Asia Minor, Fore Asia, Caucasus, Mid. Asia, Himalayas (west.), Japan, Africa.

7. **S. gebleriana** Schrenk in Fisch. et Mey. Enum. pl. nov. 1 (1841) 91; Ledeb. Fl. Ross. 1 (1842) 311; Trautv. in Bull. Soc. natur. Moscou, 32, 1 (1860) 148; Rohrb. Monogr. Silene (1868) 203; Will in J. Linn. Soc. (London) Bot. 32 (1896) 157; Maxim. Enum. pl. Mong. (1889) 90; Kryl. Fl. Zap. Sib. 5 (1931) 1066; Schischk. in Fl. SSSR, 6 (1936) 676; Opred. rast. Sr. Azii [Key to Plants of Mid. Asia] 2 (1971) 270; Claves pl. Xinjiang. 2 (1983) 261. —**Ic.**: Fl. SSSR, 6, Plate 40, fig. 2.

Described from East. Kazakhstan (Tarbagatai). Type in St.-Petersburg (LE).

In swamped sections, sands in foothills.

IIA. Junggar: *Jung. Gobi* (lower Borotal, Sept. 3, 1878—A. Reg.), *Dzhark.* (Suidun, July 16, 1877—A. Reg.).
General distribution: Aralo-Casp., Fore Balkh., Nor. Tien Shan; Mid. Asia.

8. **S. claviformis** Litw. in Trav. Mus. Bot. Ac. Sci. Petersb. 3 (1907) 99; Schischk. in Fl. SSSR, 6 (1936) 651; Opred. rast. Sr. Azii [Key to Plants of Mid. Asia] 2 (1971) 267. —*S. heptapotamica* Schischk. in Acta Inst. bot. Ac. Sci. URSS, ser. 1, 2 (1936) 147; Schischk. in Fl. SSSR, 6 (1936) 651; Fl. desert. Sin. 1 (1985) 460, tab. 168, fig. 1-2.
Described from Kazakhstan (Fore Balkh.). Type in St.-Petersburg (LE).
On sand.

IB. Kashgar: *Nor.* ("Kucha"—Fl. desert. Sin. l.c.).
IIA. Junggar: *Tarb.* ("Dachen"—Fl. desert. Sin. l.c.), *Jung. Gobi* ("Savan"—Fl. desert. Sin. l.c.).
General distribution: Fore Balkh.

9. **S. holopetala** Bge. in Ledeb. Fl. Alt. 2 (1830) 142; Ledeb. Fl. Ross. 1 (1942) 311; Kar. et Kir. in Bull. Soc. natur. Moscou, 15, 1 (1842) 166; Rohrb. Monogr. Silene (1868) 201; Maxim. Enum. pl. Mong. (1889) 90; Will. in J. Linn. Soc. (London) Bot. 32 (1896) 157; Schischk. in Fl. SSSR, 6 (1936) 677; Opred. rast. Sr. Azii [Key to Plants of Mid. Asia] 2 (1971) 270; Fl. desert Sin. 1 (1985) 457, tab. 167, fig. 2-3.
Described from East. Kazakhstan. Type in St.-Petersburg (LE). Plate 9, fig. 4.
On steppe slopes, solonetzes.

IIA. Junggar: *Zaisan* (in steppe along Kenderlyk river, Aug. 2, 1876—Pot.).
General distribution: Fore Balkh., Jung.-Tarb.

10. **S. pamirensis** (H. Winkl.) Preobr. ex Schischk. in Acta Inst. Bot. Ac. Sci. URSS, ser. 1, 2 (1936) 149; Schischk. in Fl. SSSR, 6 (1936) 670; Fl. Tadzh. 3 (1968) 563. —*S. caucasica* var. *pamirensis* H. Winkl. in Vidensk. Meddel. Natur. Foren. Kobhav. (1901) 47. —*S. karaczukuri* B. Fedtsch. in Tr. Bot. Muz. Akad. nauk. 7 (1910) 149; Schischk. in Fl. SSSR, 6 (1936) 671; Fl. Tadzh. 3 (1968) 565; Fl. Kazakhst. 3 (1960) 384; Claves pl. Xinjiang. 2 (1983) 262. — Ic.: Fl. SSSR, 6, Plate 39, fig. 4.
Described from Sinkiang. Type in St.-Petersburg (LE).
On rocks.

IIIC. Pamir (Tagdumbash-Pamir, at confluence of Kara-Chukur and Ilik-su rivers, on nor. rocks, July 17, 1901—Alekseenko, typus !).
General distribution: Mid. Asia.

11. **S. lhassana** (Will.) Majumdar in Bull. Bot. Serv. India, 15, 1-2 (1973) 42. —*Melandrium lhassanum* Will. in J. Linn. Soc. (London) Bot. 38 (1909) 406; Zhou Li-hou in Fl. Xizang. 1 (1983) 728. —Ic.: Fl. Xizang. 1, tab. 232, fig. 5-9.

Described from Tibet. Type in London (K).
On alpine meadows, scrubs, 3800-4400 m alt.

IIIB. Tibet: *South.* ("Hills above the city of Lhassa, No. 1112, 1904, Walton"—Will. l.c.; "Lhasa"—Zhou Li-hou, l.c.).
General distribution: endem.

12. S. lithophila Kar. et Kir. in Bull. Soc. natur. Moscou, 15 (1842) 167; Ledeb. Fl. Ross. 1 (1942) 776; Rohrb. Monogr. Silene (1868) 193; Schischk. in Fl. SSSR, 6 (1936) 618; Opred. rast. Sr. Azii [Key to Plants of Mid. Asia] 2 (1971) 262; Claves pl. Xinjiang. 2 (1983) 262. —**Ic.:** Fl. Kazakhst. 3, Plate 34, fig. 4.

Described from East. Kazakhstan (Jung. Ala Tau). Type in St.-Petersburg (LE). Plate 9, fig. 2.

On rocky steppe slopes, rocks.

IIA. Junggar: *Cis-Alt.* (Qinhe, No. 730, Aug. 29, 1956—Ching), *Tien Shan* (Maraltu before Muzart, Aug. 1, 1877—Fet.; Sairam, June; Talki gorge, July 25, 1877; upper Borotala, Aug. 1878; Nilki foothills, June 1879—A. Reg.; Savan area, 6 km south-east of Shichan, No. 2266, July 26, 1957—Kuan; Ketmen' mountain range, nor. slope, 3-4 km beyond Sarbushin settlement on Ili—Kalmak-kure road, steppe belt, Aug. 23, 1957—Yun. and I-F. Yuan'; hill road from Bortu to timber works in Khomote, No. 6946, Aug. 3, 1958—Lee and Chu).
General distribution: Jung.-Tarb., Nor. Tien Shan.

13. S. maximowicziana Yu. Kozhevn. stat. et nom. nov. —*S. foliosa* var. *mongolica* Maxim., Enum. pl. Mong. (1889) 91.

Described from Mongolia. Type in St.-Petersburg (LE). Plate 9, fig. 1.

IA. Mongolia: *Ordos* (Ulan-Morin river valley, Aug. 22, 1884—Pot., typus !).
General distribution: endem.

Note. This species differs from *S. foliosa* Maxim. (Prim. Fl. Amur. (1859) 53) in smaller calyx (5-6 instead of 10-11 mm), leafy short-cuspidate bracts (lanceolate in *S. foliosa*), not sticky peduncles, narrower and sparse leaves, thin yellow-coloured rachis (rhizome grey, stout in *S. foliosa*).

14. S. mongolica Maxim. Enum. pl. Mong. (1889) 88; Grub. Konsp. fl. MNR [Conspectus of Flora of Mongolian People's Republic] (1955) 131; id. Opred. rast. Mong. [Key to Plants of Mongolia] (1982) 104. —**Ic.:** Enum. pl. Mong. tab. 13, fig. 7-12; Grub. Opred. rast. Mong. [Key to Plants of Mongolia] Plate 44, fig. 205.

Described from Mongolia (Gobi). Type in St.-Petersburg (LE). Map 5.

On barren rocky steppe slopes, rocks, flanks and floors of gorges.

IA. Mongolia: *Depr. Lakes* (south-west. foothills of Dzhargalant-Ula mountain range, on rocky peak of cone-shaped hill, 1650 m, Aug. 24, 1984—Gub.), *Gobi-Alt.* (in valley of Tostu mountain, Aug. 18 1886—Pot. typus ! Tostu mountain range, south. slope, gorge of main ravine, about 2000 m, Aug. 15; Nemegetu-Nuru mountain range, on main peak at 2700 m alt., on nor. slope under rock, Aug. 8, 1948—Grub.; Khuren-Khana mountain range, along Musarin-Khundii gorge, about 1800 m alt., rocky mountain trail, Sept. 7, 1973—Grub., Dariima et al; Noyan somon, 2-3 km south of somon, on slopes of rather low cone-shaped hills with barren steppe, July 25, 1943—Yun.).

General distribution: endem.

15. **S. moorcroftiana** Wall. ex Rohrb. Monogr. Silene (1868) 129; Henders and Hume, Lahore to Yarkand (1873) 312; Edgew. et Hook. f. in Fl. Brit. India, 1 (1874) 219; Hemsl. in J. Linn. Soc. (London) Bot. 35 (1902) 170; Will. in J. Linn. Soc. (London) Bot. 38 (1909) 404; Schmid in Feddes Repert. 31 (1933) 40.

Described from Himalayas. Type in London (K).

Type of habitat not known.

IIIB Tibet: *Chang Tang* ("Keptung-La, auf granit- und gneisboden, July 29, 1927, Bosshard"—Schmid, l.c.), *South.* ("Near Rakas-Tal, from 4570 to 5180 m, 1848, Strachey, Winterbottom"—Hemsl. l.c., Williams l.c.).

General Distribution: Himalayas.

Note. It has been pointed out that Tibetan plants differ from Himalayan in greater height, slender form and number of flowers (2-3) (Hooker, l.c.).

16. **S. nana** Kar. et Kir. in Bull. Soc. natur. Moscou, 15, 1 (1842) 169; Ledeb. Fl. Ross. 1 (1842) 777; Rohrb. Monogr. Silene (1868) 157; Will. in J. Linn. Soc. (London) Bot. 32 (1898) 117; Schischk. in Fl. SSSR, 6 (1936) 682; Opred. rast. Sr. Azii [Key to Plants of Mid. Asia] 2 (1971) 272. —Ic.: SSSR, 6, Plate 41, fig. 2.

Described from Kazakhstan. Type in Moscow (MW). Isotype in St.-Petersburg (LE). Plate 9, fig. 3.

On hummocky sand.

IIA. Junggar: *Dzhark.* (Suidun, May 8, 1878—A. Reg.), *Jung. Gobi* (Paotai-Syaedi, on sand, No. 840, June 12, 1957—Kuan).

General distribution: Fore Balkh.; Fore Asia, Mid. Asia.

17. **S. noctiflora** L. Sp. pl. (1753) 419; Ledeb. Fl. Ross. 1 (1842) 314; Kar. et Kir. in Bull. Soc. natur. Moscou, 15, 1 (1842) 169., —*Melandrium noctiflorum* (L.) Fries in: Lindbl. Bot. Notiser, 10 (1842) 170; Kryl. Fl. Zap. Sib. 5 (1931) 1080; Schischk. in Fl. SSSR, 6 (1936) 712; Fl. Uzbek. 2 (1963) 396; Fl. Tadzh. 3 (1968) 592; Opred. rast. Sr. Azii [Key to Plants of Mid. Asia] 2 (1971) 281. —*Elisanthe noctiflora* Rupr. Fl. Cauc. (1869) 201; Fl. Tsentr. Sib. 1 (1979) 328.

Described from Europe. Type in London (Linn.).

Along forest borders, shrubs.

IIA. Junggar: *Cis-Alt.* (in Shara-Sume town, No. 2684, Sept. 6, 1956—Ching), *Tien Shan* (Talki gorge, July 16, 1877—A. Reg.; Ili—Chapchal, Dzhagastai, on slope, No. 3175, Aug. 8, 1957—Kuan).

General distribution: Europe, Caucasus, West. Sib.

18. **S. odoratissima** Bge. in Ledeb. Fl. Alt. 2 (1830) 148; Ledeb. Fl. Ross. 1 (1842) 312; Rohrb. Monogr. Silene (1868) 195; Kryl. Fl. Zap. Sib. 5 (1931) 1061; Schischkin Fl. SSSR, 6 (1936) 641; Opred. rast. Sr. Azii [Key to Plants of Mid. Asia] 2 (1971) 264. —*S. radians* Kar. et Kir. in Bull. Soc. natur. Moscou, 14 (1842) 40; Ledeb. Fl. Ross. 1 (1842) 777.

Described from East. Kazakhstan (Zaisan lake). Type in St.-Petersburg (LE). Plate 6, fig. 2.

On sand, sandy-rocky steppes.

IIA. **Junggar:** *Tien Shan* (Algoi, Sept. 12, 1879—A. Reg.), *Jung. Gobi* (left bank of Manas river 54 km nor.-nor.-west of Paotai state farm on Chugai road, ridgy-bedded sand, June 17; right bank of Manas river 10-15 km north of "21 Regiment" state farm, south. fringe of sand, July 11; 14-15 km east of Temirtam on Chumpazy road, saxaul hammada (rocky desert), Aug. 6; 5 km south of Paotai on road to Shikhetsze, on sand dunes, June 11—1957, Yun. et al; Manas river basin, 19 km west of Syaeda state farm, on sand, No. 904, June 14; 20 km south of Tien-Shan-laoba, No. 191, June 21; San'tszyao-chshuan, near Shikho, on sand dunes, No. 1066, June 25; "30 Regiment" state farm; in Manas gorge, on sand dunes, No. 483, July 9—1957, Kuan; 4 km south-east of Bulun-Tokhoi, on sand, June 13, 1959—Lee and Chu), *Zaisan* (Chernyi Irtysh river, left bank of Dzhelkaidar, June 9; east of Cherektas town, June 12; Sary-Dzhasyk west of Burchum river, June 14—1914, Schischk.; Alkabek, sand, Aug. 6, 1908—Fedtsch.), *Dzhark.* (Inter Suidun and Ili, May 4; fl. Talki pr. Suidun, May 7, 1878—A. Reg.).

General distribution: Aralo-Casp., Fore Balkh.; Mid. Asia.

Note. Var. *olgiana* (B. Fedtsch.) Yu. Kozhevn. in Novit. syst. pl. vasc. 22 (1985) 105 is a variety of this species. In our list of localities, it has been cited for Tien Shan.

S. orientalimongolica Ju. Kozhevn. Novit. syst. pl. vasc. 21 (1984) 68. — *Melandrium orientalimongolicum* (Yu. Kozhevn.) J.Z. Zhao in Acta sci. natur. univ. Intramong. 16, 4 (1985) 586.

Described from Mongolia (Fore Hing.). Type in St.-Petersburg (LE).

On rocks and arid mountain slopes.

Note. Though this species has been described and reported so far only from some points in Fore Hinggan (Zhao, l.c.), its find in East. Mongolia too is highly probable.

19. **S. pseudotenuis** Schischk. in Not. syst. (Leningrad), 6, 3 (1926) 6; Schischk. in Fl. SSSR, 6 (1936) 678; Ovchin. in Fl. Tadzh. 3 (1968) 569; Opred. rast. Sr. Azii [Key to Plants of Mid. Asia] 2 (1971) 271. —*S. schischkinii* Vved. in Fl. Uzb. 2 (1953) 392, p.p.

Described from Tien Shan. Type in Tomsk (TK).

On rocky steppe slopes.

IIA. **Junggar:** *Tien Shan* (midcourse of Sumbe river, cone-shaped hillocks in steppes, July 9, 1912—Sap. and Schischk.), *Jung. Alat.* (Ven'tsyuan'—Syata, forest steppe, No. 21, Aug. 14, 1957—Kuan).

General distribution: Mid. Asia.

20. **S. repens** Patr. in Pers. Syn. 1 (1805) 500; Ledeb. Fl. Ross. 1 (1842) 308; Turcz. in Bull. Soc. natur. Moscou, 15, 3 (1842) 579; Rohrb. Monogr. Silene (1868) 206; Maxim. Fl. Tangut. (1889) 82; id. Enum. pl. Mong. (1889) 91; Will. in J. Linn. Soc. (London) Bot. 32 (1896) 161; Hemsl. in J. Linn. Soc. (London) Bot. 36 (1904) 520; Kryl. Fl. Zap. Sib. 5 (1931) 1068; Hao in Bot. Jahrb. 68 (1938) 594; Walker in Contribs. U.S. Nat. Herb. 28, 4 (1941) 614; Schischk. in Fl. SSSR, 6 (1936) 654; Grub. Konsp. fl. MNR [Conspectus of Flora of Mongolian People's Republic] (1955) 131; Opred. rast. Sr. Azii [Key to Plants of Mongolia] (1982) 103; Ma Yu-chuan in Fl. Intramong. 2 (1978) 183, cum var. *angustifolia* Turcz.; Claves pl. Xinjiang. 2 (1983) 263; Fl. desert. Sin. 1 (1985) 460; Zhao in Acta sci. natur. univ. Intramong. 16, 4 (1985) 593, cum var. *angustifolia* Turcz., *latifolia* Turcz., *xilingensis* Y.Z. Zhao.
—**Ic.**: Fl. Intramong. 2, tab. 897, fig. 8-12; Fl. desert. Sin. 1, tab. 168, fig. 3-5.

Described from East. Siberia (Baikal lake). Type in Paris (P). Map 6.

On solonetzic meadows, rocky steppe slopes, fixed sand.

IA. Mongolia: *East. Mong.* (Mongolia chinensis, Bussun-tsholu, June 15, 1831— Bunge; Inter Kulussutajewsk and Dolon-Nor, 1870—Lom.; upper Kerulen, Dzho-van grand duke's site, July end, 1899—Pal.; Kulun-buir-norsk plain, Sitarkhe, June 7, 1899— Pot. and Soldatov; around Khailar town, meadow on sand plain, June 19; same site, lake bank, July 4, 1951—Sh.-S. Li et al; Shiliin-Bogdo-ula, nor. slope, around 1650 m, July 12; nor. fringe of Moltsog-els sand, about 1300 m, July 10, 1985—Gub.), *Gobi Alt.* (Gurban-Saikhan and Gurban-Bogdo—Grub. l.c.), *Alash Gobi* (Alashan mountain range, Yamato area, May 6; same site, Khote-gol area, in ravine, June 9, 1908—Czet.; 30 km east of Bayan-Khoto town, Baisa monastery, juniper thickets, 2100 m alt., July 1, 1957—Petr.), *Ordos* (Otokachi town 25 km south-east of town, near Khaoloitumdo lake, Aug. 1; sand-covered area, 20 km west of Dzhasakachi town, Aug. 16, 1957— Petr.).

IB. Kashgar: *Nor.* (south of Baichen, No. 2059, Sept. 22, 1957—Kuan).

IIA. Junggar: *Cis-Alt.* (Terekty river, July 6, 1903—Gr.-Grzh.; Qinhe-Daban', in gorge, No. 1596, Aug. 8; in Koktogoi region, No. 1816, Aug. 13, 1956—Ching.), *Tarb.* (on west. bank of Kuitun river, it poplar forest, No. 882, Oct. 12, 1957—Kuan), *Jung. Alat.* (peak of Kuzyun' pass, Aug. 2, 1908—Fedtsch.; Karaganty river, estuary of Taldy-bulak river, rubbly steppe, July 30, 1909—Sap.), *Tien Shan* (near Aksu river, June 15; Bain-gol river, near Okhotnich'ii settlement, June 28, 1893; Tumyrtyn settlement, Aug. 23, 1895—Rob.; Nilki, Kash river, June 17; Borgaty in Kash river valley, July 5; lower Arystyn, July 20— 1879, A. Reg.; descent into upper Tekes 7 km south-east of Aksu settlement, Aug, 24, 1957—Yun. et al; in Danyu region, No. 1386, July 16; 8 km south of Nyutsyuan'tszy, No. 680, July 19; Datszymyao, No. 4435, July 22; Savan area 6 km south-west of Shichan, No. 2272, July 26; east of Syata, No. 1324, Aug. 11; 30 km west of Chzhaosu, No. 3270, Aug. 12; Chzhaosu-Tekes, No. 3595, Aug. 16; 8 km south of Chzhaosu, No. 858, Aug. 13; Ulastai on bank of Kash river, No. 1745, Aug. 28; near Tyan'chi lake, No. 4349, Sept. 20; Tsitai (Guchen) area, in Magolyan village, No. 4421, Sept. 22—1957, Kuan; Urumchi-Nanshan, No. 0543, July 21, 1956—Ching).

IIIA. Qinghai: *Nanshan* ("Kokonor, Sining-fu and Umgegend, No. 795, Aug. 1930"— Hao, l.c.; 30 km east of Gunlyu, mountainous forest steppe, 3300 m, Aug. 7, 1959; 70 km south-east of Chzhan'e, Matisy temple, in mountain valley, 2600 m, July 12, 1958— Petr.; 108 km west of Sinin, 6 km west of Daudaikhe settlement, 3400 m, Aug. 5, 1959—Petr.).

General distribution: Jung.-Tarb., Tien Shan; Arct. (Asian), Europe, West. Sib., East. Sib., Far East, Nor. Mong., China (Dunbei), Korean peninsula, Japan.

Note. The variability of this species has been described by N.S. Turczaninow who distinguished within its framework α *vulgaris* with linear-lanceolate leaves and clavate calyx, β *latifolia* with lanceolate leaves and terete calyx and γ *angustifolia* with linear leaves and clavate calyx of extremely small flowers.

21. **S. sibirica** Pers. Syn. 1 (1805) 497; Ledeb. Fl. Ross. 1 (1842) 310; Kar. et Kir. in Bull. Soc. natur. Moscou, 15, 1 (1842) 166; Rohrb. Monogr. Silene (1868) 202; Will. in J. Linn. Soc. (London) Bot. 32 (1896) 157; Kryl. Fl. Zap. Sib. 5 (1931) 1065; Schischk. in. Fl. SSSR, 6 (1936) 676; Opred. rast. Sr. Azii [Key to Plants of Mid. Asia] 2 (1971) 270; Claves pl. Xinjiang. 2 (1983) 263; Fl. desert. Sin. 1 (1985) 458. —Ic.: Fl. Kazakhst. 3, Plate 36, fig. 1; Fl. desert. Sin. 1, tab. 167, fig. 1.

Described from Siberia. Type in London (Linn.).

On steppe slopes, sand.

IIA. **Junggar:** *Cis-Alt.* (south of Koktogoi, in Ukagou region, on arid steppe, No. 1732, Aug. 11, 1956—Ching), **Zaisan** (Chernyi Irtysh river, left bank, Maikain area, on hummocky sand, June 7, 1914—Schischk.).

General distribution: Aralo-Casp., Fore Balkh.; West. Sib. (Altay), Europe.

22. **S. subcretacea** Will. in J. Linn. Soc. (London) Bot. 38 (1907-1909) 404; Majumdar in Bull. Bot. Surv. India, 15, 1-2 (1973) 39; Zhou Li-hou in Fl. Xizang. 1 (1983) 732.

Described from Himalayas. Type in London (K). Isotype in St.-Petersburg (LE). Plate 6, fig. 3.

On alpine meadows, 4000-5000 m alt.

IIIB. **Tibet:** *South.* ("Gyangtse, No. 105, 1108, Walton"—Will. l.c.; "Tszyantsze, Lankatsza, Chzhunba"—Zhou Li-hou, l.c.).

General distribution: Himalayas (east.).

Note. This species differs from *S. cretacea* Fisch. in narrower green (not grey-green) leaves, a different structure of lower part of plant (highly characteristic of *S. cretacea* as a result of its growing on chalky substrates) and orbicular (not cuspidate) calyx teeth.

23. **S. tatarica** (L.) Pers. Syn. 1 (1805) 497; Ledeb. Fl. Ross. 1 (1842) 312; Rohrb. Monogr. Silene (1868) 184; Kryl. Fl. Zap. Sib. 5 (1931) 1057; Schischk. in Fl. SSSR, 6 (1936) 619. —*Cucubalus tataricus* L. Sp. pl. (1753) 415.

Described from "Tataria". Type in London (Linn.).

On meadowy steppes.

IIA. **Junggar:** *Tien Shan* (Khorgos river valley nor.-west of Kul'dzha, Aug. 1881—A. Reg.).

General distribution: Europe, Caucasus.

24. **S. tenuis** Willd. Enum. pl. hort. Berol. (1809) 474; Kar. et Kir. in Bull. Soc. natur. Moscou, 15, 1 (1842) 166; Turcz. in Bull. Soc. natur. Moscou, 15, 3 (1842) 577; Rohrb. Monogr. Silene (1868) 186; Edgew. et Hook. f. in Fl. Brit. India, 1 (1875) 219; Maxim. Enum. pl. Mong. (1889) 89; Will. in J. Linn. Soc. (London) Bot. 34 (1898-1900) 429; Hemsl. in J. Linn. Soc. (London) Bot. 36 (1904) 521; Pamp. Fl. Caracor. (1930) 106; Rehder and Kobuski in J. Arn. Arb. 14 (1933) 9; Hao in Bot. Jahrb. 68 (1938) 594; Walker in Contribs. U.S. Nat. Herb. 28, 4 (1941) 614; Zhou Li-hou in Fl. Xizang. 1 (1983) 734. —*S. jenisseensis* Willd. Enum. pl. horti Berol. (1809) 153; Schischk. in Fl. SSSR, 6 (1936) 627; Grub. Konsp. fl. MNR [Conspectus of Flora of Mongolian People's Republic] (1955) 131; id. Opred. rast. Mong. [Key to Plants of Mongolia] (1982) 103; Claves pl. Xinjiang. 2 (1983) 261; Opred. rast. Tuv. ASSR [Key to Plants of Tuva Autonomous Soviet Socialist Republic] (1984) 59; Fl. desert. Sin. 1 (1985) 459; Zhao in. Acta sci. natur. univ. Intramong. 16, 4 (1985) 595. —*S. graminifolia* Otth. in DC. Prodr. 1 (1824) 368; Ledeb. Fl. Ross. 1 (1842) 307; Kryl. Fl. Zap. Sib. 5 (1931) 1058; Schischk. in Fl. SSSR, 6 (1936) 625; Grub. Konsp. fl. MNR [Conspectus of Flora of Mongolian People's Republic] (1955) 131; Ikonnik. Opred. rast. Pamira [Key to Plants of Pamir] (1963) 108; Zhao, l.c.; Opred. rast. Sr. Azii [Key to Plants of Mid. Asia] 2 (1971) 262. —*S. chamarensis* Turcz. l.c.; Kryl. l.c. 1059; Grub. l.c. 1955, 1982. —*S. tenuis* subsp. *chamarensis* (Turcz.) Ju. Kozhevn. in Novit. syst. pl. vasc. 22 (1985) 110. —*S. iche-bogdo* Grub. in Not. syst. (Leningrad), 17 (1955) 13; Grub. Konsp. fl. MNR [Conspectus of Flora of Mongolian People's Republic] (1955) 131; id. Opred. rast. Mong. [Key to Plants of Mongolia] (1982) 103. —*S. tenuis* subsp. *iche-bogdo* (Grub.) Yu. Kozhevn. l.c.. —**Ic.**: Grub. Opred. rast. Mong. [Key to Plants of Mongolia] Plate 44, fig. 206; Fl. Xizang. 1, tab. 237.

Described from specimens grown in Berlin from Fore Baikal seeds. Type in Berlin (B).

On rocky steppe slopes, on rocks, alpine meadows, larch groves.

IA. Mongolia: *Khobd., Mong. Alt., East. Mong., Depr. Lakes, Gobi-Alt., East. Gobi, Ordos.*

IB. Kashgar: *Nor.* (Uchturfan, Karagailik gorge, June 18, 1908—Divn.).

IC. Qaidam: *South* (nor. slope of Burkhan-Budda mountain range, Khatu gorge, on precipices and rocks, July 12, 1901—Lad.), *mount.* (Dulan-khit temple, in spruce and juniper forests, Aug. 12, 1901—Lad.).

IIA. Junggar: *Cis-Alt.* (Koktogoi, No. 1983, Aug. 17; Qinhe, between Kun'tai village and Chzhunkhaitsza, No. 1154, Aug. 5, 1956—Ching), *Tien Shan.*

IIIA. Qinghai: *Nanshan* (in trough toward Tetung river, July 1; Tetung river valley, around ploughed fields, rarely on grassy descent of mountain, Aug. 1; in alpine belt of mountain nor. of Tetung river, Aug. 28, 1872; on plateau around Kuku-nor, July 19, 1880—Przew.; nor. of Gan'-Chzhou town, June 26, 1875—Pias.; South Kukunor mountain range, Bain-gol river, in meadow, June 26, 1894—Rob.; 89 km west of Sinin, pass, Aug. 5, 1959—Petr.), *Amdo* ("Kokonor, auf dem Hochen-Plateau Da-ho-ba, No. 1059, Aug. 28 1930"—Hao, l.c.; "Radja and Yellow River gorges, No. 14209"—Rehder and Kobuski, l.c.).

IIIB. Tibet: *Weitzan* (Dzhagyn-gol river, No. 295, data lacking—Lad.).

IIIC. **Pamir** (Tagdumbash-Pamir, in Pistan gorge of Sary-Kol mountain range, among rocks, July 15, 1920—Alekseenko).

General distribution: Aralo-Casp., Fore Balkh., Jung.-Tarb., Nor. and Cent. Tien Shan; Europe, Mid. Asia, West. Sib. (Altay), East. Sib., Nor. Mong., China (Nor., Nor.-West.), Himalayas (west., Kashmir).

25. **S. viscosa** (L.) Pers. Syn. 1 (1805) 497; Ledeb. Fl. Ross. 1 (1842) 313; Maxim. Enum. pl. Mong. (1889) 90. —*Cucubalus viscosus* L. Sp. pl. (1753) 414. —*Melandrium viscosum* (L.) Cel. in Lotos, 18 (1868); Schischk. in Fl. SSSR, 6 (1936) 710; Grub. Konsp. fl. MNR [Conspectus of Flora of Mongolian People's Republic] (1955) 133; Opred. rast. Sr. Azii [Key to Plants of Mid. Asia] 2 (1971) 280; Grub. Opred. rast. Mong. [Key to Plants of Mongolia] (1982) 104. —*M. suaveolens* (Kar. et Kir.) Schischk. in Fl. URSS, 6 (1936) 711; Grub. l.c. 133; Grub. Opred. rast. Mong. [Key to Plants of Mongolia] (1982) 104; Claves pl. Xinjiang. 2 (1983) 258. —*Silene suaveolens* Kar. et Kir. in Bull. Soc. natur. Moscou, 15 (1842) 168; Ledeb. Fl. Ross. 1 (1842) 777. —*Melandrium quadrilobum* (Turcz.) Schischk. l.c. 711; Grub. Konsp. fl. MNR [Conspectus of Flora of People's Republic of Mongolia] (1955) 133; id. Opred. rast. Mong. [Key to Plants of Mongolia] (1982) 104. —*Silene quadriloba* Turcz. ex Kar. et Kir. l.c. 167. —**Ic.:** Fl. Tadzh. 3. Plate 91, fig. 1-5; Fl. SSSR, 6, Plate 43, fig. 6.

Described from Europe. Type in London (Linn.). Plate 5, fig. 4 and 5.

On steppe slopes, meadows.

IA. Mongolia: *Depr. Lakes* (nor.-east. part of Bayan-nur lake, vicinity of Borig-del' sand, July 13, 1973—Banzragch et al; south. extremity of Ubsu-nur basin, Borig-Del' sand, Tsagan-del' area, July 21, 1971—Grub., Ulzij. et al; granite conical hillock 3 km south of ulangom on road to Khobdo, gorge, July 27; Borig-del' sand, 4-5 km south-east of Baga-nur lake, sand dunes, July 25, 1945—Yun.; 10 km east of Ulangom, barren hillock, 1100 m, Aug. 25, 1979; *Colocasia* (Taro) desert 85 km east of Ulangom town, about 970 m, Sept. 3, 1984—Gub.), *Mong. Alt.* (upper Tsagan-gol, rocks and coastal meadow, July 3, 1905—Sap.; 3-4 km south-east of Tamchi-nur lake, July 17, 1947—Yun.; Buyantu river basin, west. spur of Bugu-ula on road through Dzhangyz-Agach to Tushintu-daba, 5 km nor. of pass, 2260 m alt., July 2, 1971—Grub., Ulzij. et al. Baitag-Bogdo mountain range, Budun-Khargaityn-gol basin, subalpine spiny-hummocky area, 2100 m, July 28, 1979—Gub.).

IIA. Junggar: *Cis-Alt.* (2 km south-south-east of Shara-Sume settlement, on road to Shipati (through Irtysh), rocky desert steppe, July 7; 30 km nor. of Koktogoi, right bank of Kairta river, Kuidyn river valley, forest belt, July 15—1959, Yun., I.-F. Yuan'; Qinhe, on slope, No. 1289, Aug. 2, 1956—Ching), *Jung. Alat.* (in Toli area, Aug. 7, 1957—Kuan), *Tien Shan* (in steppe on Bain-gol river (upper Terekta river), June 27, 1893—Rob.; Dzhagastai, 1200-1500 m, Aug. 9, 1877—A. Reg.; on Kunges river, in forest, June 27, 1877—Przew.; east of Aksu on road to Pshaksu, No. 1595, Aug. 15, 1957—Kuan), *Jung. Gobi* (on Savan town to Paotai road, No. 736, June 10, 1957—Kuan), *Dzhark.* (Aktyube around Kul'dzha, 900 m, May 13, 1877—A. Reg.).

General distribution: Aralo-Casp., Fore Balkh., Jung.-Tarb., Tien Shan; Europe, Caucasus, Mid. Asia, West. and East. Sib., Nor. Mong. (Hang.).

Note. *S. griffithii* Boiss. with longer calyx (18-25 mm), distributed in west. Himalayas (Hooker, 1875), probably belongs to this species.

26. **S. vulgaris** (Moench) Garcke, Fl. Nord-Mittel-Deutschl., ed. 9 (1869) 64; Grub. Opred rast. Mong. [Key to Plants of Mongolia] (1982) 103; Claves pl. Xinjiang. 2 (1983) 262. —*Cucubalus behen* L. Sp. pl. (1753) 415. —*C. latifolius* Mill. Gard. Dict. ed. 8 (1768) No. 2. —*C. venosus* Gilib. Fl. Lithuan. 2 (1781) 165. —*Silene latifolia* (Mill.) Britt. et Rendle, List Brit. Seed-Plants (1907) 5, non Poir.; Schischk. in Fl. SSSR, 6 (1936) 596. —*Behen vulgaris* Moench, Meth. (1794) 709; Fl. Kazakhst. 3 (1960) 368. —*Cucubalus inflatus* Salisb. Prodr. (1796) 302. —*Silene cucubalus* Wib. Prim. Fl. Werth. (1799) 241; Rohrb. Monogr. Silene (1868) 84; Catal. Nepal. vasc. pl. (1985) 47. —*S. inflata* Smith, Fl. Brit. (1800) 467; Ledeb. Fl. Ross. 1 (1842) 304; Kar. et Kir. in Bull. Soc. natur. Moscou,15, 1 (1842) 166; Turcz. in Bull. Soc. natur. Moscou, 15, 3 (1842) 573; Edgew. et Hook. f. in Fl. Brit. Ind. 1 (1875) 218; *S. inflata* β *litoralis* Rupr. Fl. Ingr. 1 (1860) 159. —*S. wallichiana* Klotsch in Klotsch et Garcke, Bot. Ergebn. Reise pr. Waldem. Preuss. (1862) 139, tab. 30; Claves pl. Xinjiang. 2 (1983) 264; Vved. in Fl. Uzbek. 2 (1863) 379. —*S. venosa* (Gilib.) Aschers. Fl. Prov. Brandenb. 2 (1864) 23; Pamp. Fl. Carac. (1930) 106; Grub. Konsp. fl. MNR [Conspectus of Flora of Mongolian People's Republic] (1955) 132; Opred. rast. Sr. Azii [Key to Plants of Mid. Asia] 2 (1971) 259; Catal. Nepal. vasc. pl. (1976) 47; Zhao in Acta sci. natur. univ. Intramoong. 16, 4 (1985) 592. —*Oberna wallichiana* Ikonnik. in Novit. syst. pl. vasc. 13 (1973) 120; Ikonnik. Opred. vyssh. rast. Badakhsh. [Key to Higher Plants of Badakhsh.] (1979) 151. —*Behenantha behen* (L.) Ikonnik. l.c. 15 (1975) 198. —*B. commutata* (Guss.) Ikonnik. l.c. 198. —*B. commutata* (Guss.) Ikonnik. l.c. 198. —*B. litoralis* (Rupr.) Ikonnik. l.c. 199. —*B. uniflora* (Roth) Ikonnik. l.c. —*B. wallichiana* (Klotzsch) Ikonnik. l.c. —**Ic.**: Fl. Uzbek., 3, Plate 36, fig. 1; Fl. SSSR, 6, Plate 35, fig. 4.

Described from Europe. Type in London (Linn.).

On meadows, along rivers, on garbage, roadsides.

IIA. Junggar: *Cis-Alt.* (Qinhe, No. 899, Aug. 2; Shara-Sume, No. 2403, Aug. 26— 1956, Ching; 30 km north of Koktogoi, right bank of Kairta river, Kuidyn river valley, forest belt, Aug. 15, 1959—Yun., I.-F. Yuan'), *Jung. Alat.* (in Toli area, No. 1058 and 2522, Aug. 6, 15 km nor.-west of Ven'tsyuan' lake, No. 1637, Aug. 29, 1957—Kuan), *Tien Shan* (Sharysu river, June 26; Talki, July 18; Sairam, July 20; Tekes river valley, June 23; Dzhagastai, Aug. 8, 1878; around Nilki, June 8 and 16; on Kapchagai-Borgata road, July 6, 1879—A. Reg., on Sairam lake, July 23, 1878—Fet.; left bank of Manas river, Ulan-Usu river valley, 1 km downstream of Koisu confluence, spruce grove, Aug. 17; Boro-Khoro mountain range, 6 km south of N. Ortai settlement on road to Kul'dzha from Sairam-nor, lower part of forest belt, Aug. 19; upper stream of Tekes river, 4-5 km south-east of Aksu settlement, Aug. 24; same site, forest belt and meadow, Aug. 24— 1957, Yun. et al; Koisu river, No. 93, July 17; in Savan region, in forest, No. 1670, July 19; Savan area, Dat-szymyao village, No. 1710, July 22; south of Shichan town, on shaded slope, No. 823, July 23; Savan area, 6 km south-west of Shichan, No. 2781, July 26; Ili-Chapchal, Dzhagastai, No. 3165, Aug. 8; Chzhaosu area, from Syada to Ven'tsyuan', in forest, No. 3434, Aug. 13; from Chzhaosu 8 km northward, No. 931, Aug. 15; Sin'shan', Nanshan, on nor. slope, No. 1160, Aug. 22; Ven'tsyuan' area, No. 4626, Aug. 25; on Tyan'chi lake, No. 4321, Sept. 19—1957, Kuan).

General distribution: Aralo-Casp., Fore Balkh., Jung.-Tarb., Tien Shan; Arct. (Europ.), Europe, Mediterr., Balk.-Asia Minor, Fore Asia, Caucasus, Mid. Asia, West. and East. Sib., Nor. Mong. (Hent.), Himalayas, Afr. (nor.).

27. **S. waltoni** Will. in J. Linn. Soc. (London) Bot. 38 (1909) 404; Zhou Li-hou in Fl. Xizang. 1 (1983) 733. —Ic.: Fl. Xizang. tab. 236.

Described from Tibet. Type in London (K). Map 6.

On alpine meadows, 3000-4700 m alt.

IIIB. Tibet: *South.* ("Gyangtse, 1904, No. 1105, Walton"—Will. l.c.; "Lhasa, Lankatsza, Nan'mulin'"—Zhou Li-hou, l.c.).

28. **S. wolgensis** (Willd.) Bess. ex Spreng. Ind. sem. Horti. Halens. (1818) 7; Otth. in DC. Prodr. 1 (1824) 370; Boiss. Fl. Or. 1 (1857) 607; Kryl. Fl. Zap. Sib. 5 (1931) 1063; Schischk. in Fl. SSSR, 6 (1936) 685; Opred. rast. Sr. Azii [Key to Plants of Mid. Asia] 2 (1971) 273; Fl. desert. Sin. 1 (1985) 457. —*S. densiflora* D'Urv. in Mem. Soc. Linn. Paris, 1 (1822) 303; Schischk. in Fl. SSSR, 6 (1936) 686; Fl. Kazakhst. 3 (1960) 390; Claves pl. Xinjiang. 2 (1983) 260. —*S. otites* var. *wolgensis* Trautv. in Bull. Soc. natur. Moscou, 32, 1 (1860) 147; Rohrb. Monogr. Silene (1868) 201. —*S. cyri* Schischk. in Fl. Tiphlis, 1 (1925) 202; Schischk. in Fl. SSSR, 6 (1936) 688; Opred. rast. Sr. Azii [Key to Plants of Mid. Asia] l.c.; Fl. desert. Sin. 1 (1985) 455. —Ic.: Fl. SSSR, 6, Plate 41, fig. 1; Fl. desert. Sin. 1, tab, 166, fig. 1-4.

Described from Volga. Type in Berlin (B).

On steppe slopes, shrubs and groves along rivers, on rocks.

IIA. Junggar: *Cis-Alt.* (Kairta river valley, nor. slope, Aug. 2, 1906—Sap.; from Koktogoi in south, around Ukagou, No. 1729, Aug. 11; vicinity of Koktogoi town, No. 2142, Aug. 18, 1956—Ching; 20 km nor.-west of Shara-Sume, midmountain belt, July 7, 1959—Yun., I.-F. Yuan; 8 km north of Barbagai, on terrace 1 of Khobuk river, July 8, 1959—Lee and Chu), *Jung. Alat.* (ascent to Kuzyun' pass, rocky site, Aug. 2, 1908—Fedtsch.), *Jung. Gobi* (on way to Temirtau, in desert, No. 2968, Aug. 15, 1957—Kuan), *Zaisan* (Kaba river around Kaba village, tugai, June 16, 1914—Schischk.).

IIIA. Qinghai: *Nanshan* (33 km from Sinin, rocky slopes, Aug. 5, 1959—Petr.).

General distribution: Aralo-Casp., Fore Balkh., Jung.-Tarb., Tien Shan; Europe, Caucasus, Balk.-Asia Minor, Mid. Asia, West. and East. Sib.

Note. This species does not hybridize at all with *S. otites* (L.) Wibel, suggesting its adequate independence (Chater, Walters, l.c.). *S. cyri* can be differentiated only as an intraspecific category as its difference lies only in longer (6-9 mm) capsule.

15. Lychnis L.

Sp. pl. (1753) 436

1. Leaves lanceolate or linear-lanceolate, 1-4 cm long. Flowers on pedicels as long as calyx or slightly longer; calyx 6-8 mm long, with blunt teeth; petals white or pink .... 2. **L. sibirica** L.

+ Leaves ovoid-lanceolate, 2-8 cm long. Flowers subsessile; calyx 15-18 mm long, with sharp teeth; petals bright-red ....
.... 1. **L. chalcedonica** L.

1. **L. chalcedonica** L. Sp. pl. (1753) 436; Ledeb. Fl. Ross. 1 (1842) 330; Gorschk. in Fl. SSSR, 6 (1936) 696; Fl. Kazakhst. 3 (1960) 396; Opred. rast. Sr. Azii [Key to Plants of Mid. Asia] 2 (1971) 278; Claves pl. Xinjiang. 2 (1983) 256; Fl. desert. Sin. 1 (1985) 463. —Ic.: Fl. SSSR, 6, Plate 45, fig. 4.
Described from Europe. Type in London (Linn.).
On moist meadows, scrubs, lowlands.

IIA. **Junggar:** *Zaisan* ("Burchum"—Claves pl. Xinjiang. l.c.).
General distribution: Aralo-Casp., Fore Balkh., Tien Shan; Europe, West. Sib., East. Sib.

2. **L. sibirica** L. Sp. pl. (1753) 437; Kryl. Fl. Zap. Sib. 5 (1931) 1072; Gorschk. in Fl. SSSR, 6 (1936) 693; Grub. Konsp. fl. MNR [Conspectus of Flora of Mongolian People's Republic] (1955) 132; id. Opred. rast. Mong. [Key to Plants of Mongolia] (1982) 104; Opred. rast. Tuv. ASSR [Key to Plants of Tuva Autonomous Soviet Socialist Republic] (1984) 60. —Ic.: Fl. SSSR, 6, Plate 42, fig. 1; Grub. Opred. rast. Mong. [Key to Plants of Mongolia] Plate 45, fig. 209.
Described from Siberia. Type in London (Linn.).
On rubbly slopes, rocks.

IA. **Mongolia:** *East.-Mong.* (Grub. l.c.).
General distribution: Arct., Europe, West. Sib., East. Sib., Far East, Nor. Mong. (Fore Hubs., Hent., Hang., Mong.-Daur.).

16. **Melandrium** Roehl.

Deutschl. Fl. ed. 2, 2 (1812) 274. —*Melandryum* Reichb. Handb.
(1837) 298. —*Melandrum* Blytt, Georges Fl. 3 (1876) 1068

1. Rhizome stout, lobed. Plant weak, soft, bedded, with linear leaves. Peduncles not longer than leaves....6. **M. neocaespitosum** J.W. Tsui.
+   Rhizome short, not lobed, sometimes funiform. Plants strong, generally not bedded, with various leaf shapes but not linear. Peduncles longer than leaves .... 2.
2. Petals entire, with glabrous claw, slightly longer than calyx ....
.... 5. **M. integripetalum** L.H. Zhou.
+   Petals lobed, with scales, as long as calyx or longer.................. 3.
3. Pubescence of jointed and simple white eglandular hairs (sometimes hairs glandular but nevertheless light-coloured with very small heads). Calyx 10-12 mm long (in fruit up to 18 mm), with green (rarely purple in upper part) nerves and sharp narrow-scarious teeth .... 3. **M. brachypetalum** (Horn.) Fenzl.

+   Pubescence of glandular hairs with wine-coloured septa of segments admixed with unjointed hairs. Calyx more than 12 mm long, with wine-coloured nerves and blunt narrow- or broad-scarious teeth .... 4.

4.   Radical leaves 8-15 mm broad. Flowers in loose racemes. Seeds not winged.. 5.

+   Radical leaves narrower than 8 mm. Flowers solitary, rarely 2-4. Seeds winged .... 2. **M. apetalum** (L.) Fenzl.

5.   Plant 30-40 cm tall, forming beds. Cauline leaves 5-8 cm long, considerably longer than radical leaves. Petals divided into narrow lobes, with squarish scales. Capsule elongated ....

.... **M. multifurcatum** (C.L. Tang) Yu. Kozhevn.

+   Plant under 30 cm tall, not forming beds (though *M. alaschanicum* forms a typical caudex characteristic of xerophytes). Cauline leaves shorter than 5 cm and considerably smaller than radical leaves. Petals divided into broad (middle) and narrow (lateral) lobes with small orbicular scales. Capsule orbicular or broad-elliptical .... 6.

6.   Petals longer than calyx, corolla rotate. Radical leaves spatulate. Seeds about 1.8 mm in diam..

.... 1. **M. alaschanicum** (Maxim.) J.Z. Zhao.

+   Petals as long as calyx or slightly longer, corolla not rotate. Radical leaves broad-lanceolate. Seeds about 0.8 mm in diam ....

.... 4. **M. glandulosum** (Maxim.) Will.

1. **M. alaschanicum** (Maxim.) J.Z. Zhao in Acta sci. natur. univ. Intramong. 16, 4 (1985) 588. —*Lychnis alaschanica* Maxim. in Mel. biol. 10 (1880) 577; id. Enum. pl. Mong. (1889) 92. —Ic.: Enum. pl. Mong., Plate 6, fig. 1-8.

Described from Mongolia. Type in St.-Petersburg (LE). Plate 2, fig. 6. Map 5.

In crevices on wet sites.

    IA. **Mongolia:** *Alash. Gobi* (on west. slope in central Alashan, in gorges on moist humus, July 10, 1873—Przew., typus !).
    General distribution: endemic.

2. **M. apetalum** (L.) Fenzl in Ledeb. Fl. Ross. 1 (1842) 326; Hedin, S. Tibet (1922) 84; Hand.-Mazz. Symb. Sin. 7 (1929) 208; Pamp. Fl. Carac. (1930) 107; Kryl. Fl. Zap. Sib. 5 (1931) 1075; Schischk. in Fl. SSSR, 6 (1936) 716; Hao in Bot. Jahrb. 68 (1938) 595; Walker in Contribs. U.S. Nat. Herb. 28, 4 (1941) 613; Grub. Konsp. fl. MNR [Conspectus of Flora of People's Republic of Mongolia] (1955) 132; Opred. rast. Sr. Azii [Key to Plants of Mid. Asia] 2 (1971) 282; Grub. Opred. rast. Mong. [Key to Plants of Mongolia] (1982) 104; Ikonnik. Opred. rast. Pamira [Key to Plants of Pamir] (1963) 109; Zhou

Li-hou in Fl. Xizang. 1 (1983) 713; Claves pl. Xinjiang. 2 (1983) 257; Zhao in Acta sci. natur. univ. Intramong. 16, 4 (1985) 584. —*M. nigrescens* (Edgew.) Will. in J. Linn. Soc. (London) Bot. 38 (1909) 405; Pamp. l.c. 107; Hand.-Mazz. Symb. Sin. 7 (1929) 208; Rehder and Kobuski in J. Arn. Arb. 14 (1933) 9; Hao in Bot. Jahrb. 68 (1938) 595. —*M. auripetalum* J.Z. Zhao et Ma in Acta Phytotax. Sinica, 27, 3 (1989) 225. —*M. verrucoso-alatum* J.Z. Zhao et Ma, ibid. 227. —*Lychnis apetala* L. Sp. pl. (1753) 437; Kar. et Kir. in Bull. Soc. natur. Moscou, 15, 1 (1842) 170; Turcz. in Bull. Soc. natur. Moscou, 15, 3 (1842) 584; Henders and Hume, Lahore to Yarkand (1873) 312; Edgew. et Hook. f. in Fl. Brit. India, 1 (1875) 222; Maxim. Enum. pl. Mong. (1889) 93; id. Fl. Tangut. (1889) 82; Deasy, In Tibet and Chinese Turk. (1901) 400; Hemsl. in J. Linn. Soc. (London) Bot. 35 (1902) 169; Stewart in Bull. Torrey Bot. Club (1916) 632. —*L. nigrescens* Edgew. in Hook. f. Fl. Brit. India, 1 (1875) 223. —*L. macrorhiza* Royle, Ill. Bot. Himal. 1 (1839) 80; Edgew. et Hook. f. in Fl. Brit. India, 1 (1875) 223; Hemsl. in J. Linn. Soc. (London) Bot. 35 (1902) 169. —*Gastrolychnis apetala* (L.) Tolm. et Kozh. in Arkt. fl. SSSR, 6 (1971) 113. —*Silene nigrescens* (Edgew.) Majumdar in Bull. Bot. Surv. India, 15, 1 (1973) 44; Catal. Nepal vasc. pl. (1976) 47. —Ic.: Fl. SSSR, 6, Plate 44, fig. 2; Grub. Opred. rast. Mong. [Key to Plants of Mongolia] Plate 45, fig. 210; Fl. Xizang. 1, tab. 228, fig. 5-9.

Described from Lapland and Siberia. Type in London (Linn.).

In swampy sections, wet and alpine meadows, forests, along river banks and shoals, up to 5000 m.

IA. Mongolia: *Khobd., Mong. Alt., Depr. Lakes, Gobi-Alt.*

IIA. Junggar: *Jung. Alat., Tien Shan.*

IIIA. Qinghai: *Nanshan* (in alpine belt, July 23, 1879; on Rako-gol river, July 22, 1880—Przew.; Mon'yuan', moraine in sources of Ganshig river, source of Peishikhe river, 3900-4300 m, Aug. 18; on bank of Peishikhe river, 3900-4300 m, Aug. 18—1958, Dolgushin), *Amdo* (on bank of Baga-Gorgi river, May 26, 1880—Przew.; "grasslands between Labrang and Yellow River, No. 14476, 14500"—Rehder and Kobuski, l.c.).

IIIB. Tibet: *Chang Tang* (nor. slope of Russky mountain range, June 2, 1890—Rob.; Keriya, in Kyuk-Egil' gorge, along descent on alpine meadows, July 2, Aug. 2, 5, 1885—Przew.; "Mandarlic, 3437 m, medio July; *Chimen-tagh, Kar-yajlak-sai, camp X, 3484 m, July 21, 1900; Tibet, without locality"—Hedin, l.c.; *"Kiam, 5100 m, Aug. 11; von Klam nach Lumkang, 5100-5200 m, Aug. 12; *Lacarpo am Lanak-La, 5400 m, Aug. 13; Tschu-sang-po am Lanak-La, 5450 m, Aug. 13; *Aksai-Chin, about 5000 m, Sept. 5, 1927, Bosshard"—Schmid, l.c.; "Ban'ge, Shuankhu"—Zhou Li-hou, l.c.), *Weitzan* (south. slope of Burkhan-Budda mountain range, on alpine meadows, July 12; on Razboinich'ei river, July 26; on bank of By-chyu river, July 8—1884, Przew.; Dzhagyn-gol river, 4150 m alt., July 1; Amnen-Kor mountain range, south. slope, June 10, 1900; same site, nor. slope, 4150-4200 m alt., June 5, 1901—Lad.; "Kokonor, Amne Matchin, auf den Abhangen, No. 1096, Sept. 3. 1930"—Hao, l.c.; "An'do"—Zhou Li-hou, l.c.), *South.* ("Gooring valley, about 4950 m, Littledale"—Hemsl. l.c.; "Chzhunba, Lhasa, *Chzhada, *Pulan'"—Zhou Li-hou, l.c.).

IIIC. Pamir ("Mus-tagh-ata, left old moraine of the Korumde glacier, 4367 m, July 27, 1894"—Hedin, l.c.; morainic waterdividing region between Atrakyr and Tyuzutek rivers, mossy tundra at 4500-5000 m, July 20; on Piyakien brook at 3600-4000 m alt.,

July 2; on Mia river gorge at 4000 m alt., July 21; upper Kashka-su river, below moraine, at about 3500 m alt., July 5; Tas-Pestalyk area at 4000-5000 m alt., July 25; Goo-dzhiro river, at 4500-5500 m alt., July 27; Shor-luk river gorge at 4000-5500 m alt., July 28—1942, Serp.).

General distribution: Jung.-Tarb., Nor. and Cent. Tien Shan, East. Pam.; Arct., Europe, West. Sib. (Altay), East. Sib.; Far East, Nor. Mong., Himalayas, Nor. America.

Note. This species exhibits considerable polymorphism. Rohrbach very early distinguished var. *himalayense* Rohrb. in Linnaea, 36 (1869-70) 220; Schmid in Feddes Repert. 31 (1933) 40; Zhou Li-hou, l.c. 714. —*Lychnis himalayensis* Edgew. in Hook. f. l.c. 223; Stewart, l.c. Catal. Nepal vasc. pl. (1976) 46. —*Melandrium himalayense* (Rohrb.) Y.Z. Zhao in Acta sci. natur. univ. Intramong. 16, 4 (1985) 585.

This variety differs from type foremost in size. Plant 20-40 cm tall; leaves long, narrow; calyx 6-8 mm long. It has been asterisked (*) in our list. Edgeworth (l.c.) assumed that *Lychnis himalayensis* could only be a variety of *L. apetala* L. although he promoted the rank of variety assigned by Rohrbach to species and transferred it to another genus. It should be pointed out that tall plants with narrow long cauline leaves occur not only in Himalayas but all over Mongolia and Siberia but their calyx may be of normal size, i.e. 13-16 mm long or more. Such plants obviously do not belong to var. *himalayense*.

Edgeworth and Hooker (l.c.) also recognised var. *pallida* having very pale calyx with green indistinct nerves and usually 2 flowers on peduncles. The taxonomic significance of these characteristics is doubtful as there is no correlation between them. Large plants of *Melandrium apetala* often have 2-3 flowers on the same stem but the size of calyx, delineation and coloration of their nerves as well as their pubescence vary widely. In many cases, plants have altogether glabrous calyx while in other cases eglandular hairs predominate in pubescence and in some others pubescence consists almost completely of glandular multicellular hairs. Coronal scales on petals may be present or absent. Filaments glabrous or pubescent. All these characteristics are not of diagnostic importance as was accepted by some botanists. Intraspecific variability is also demonstrated by distinct "auricles" on petals, pubescence of stem, abundance of radical leaves, development of seed wing and its sculpture; according to some patterns of latter, *M. auripetalum* and *M. verrucoso-alatum* were recently differentiated.

3. **M. brachypetalum** (Horn.) Fenzl in Ledeb. Fl. Ross. 1 (1842) 326; Turcz. in Bull. Soc. natur. Moscou, 15, 3 (1842) 585; Hand.-Mazz. Symb. Sin. 7 (1929) 205; Schischk. in Fl. SSSR, 6 (1936) 722; Grub. Konsp. fl. MNR [Conspectus of Flora of Mongolian People's Republic] (1955) 132; Opred. rast. Sr. Azii [Key to Plants of Mid. Asia] 2 (1971) 282; Grub. Opred. rast. Mong. [Key to Plants of Mongolia] (1982) 104; Ma Yu-chuan in Fl. Intramong. 2 (1978) 181; Zhao in Acta sci. natur. univ. Intramong. 16, 4 (1985) 588.

—*Lychnis brachypetala* Horn. Hort. Hafn. Suppl. (1819) 51; Edgew. et Hook. f. in Fl. Brit. India, 1 (1875) 223; Maxim. Enum. pl. Mong. (1889) 93. — *Gastrolychnis brachypetala* (Horn.) Tolm. et Kozh. in Arkt. fl. SSSR, 6 (1971) 109. —Ic.: Fl. SSSR, 6, Plate 44, fig. 7; Fl. Intramong. 2, tab. 96, fig. 7-11.

Described from specimens grown from seeds of unknown origin. Type in Copenhagen (C).

Along steppe and meadow slopes, on rocks.

IA. **Mongolia:** *Khobd.* (along south. source of Kharkhira river, July 23, 1879—Pot.; Turgen river valley, July 8, 1973—Banzragch), *Mong. Alt.* (Daingol lake, steppe slopes, July 5, 1906—Sap.; Tolbo-Kungei-nuru mountain range, upper mountain belt, Aug. 5, 1945. Adzhi-Bogdo mountain range, ascent to Burgastyin-dava pass in upper Inder-tiin-gol, mountain steppe, Aug. 6, 1947—Yun.; upper Bulugun river, Artelin-sala creek valley on south. slope of Shara-Khalusyn-Orai mountain range, 3 km beyond Kudzhurtu settlement, July 3, 1971—Grub., Ulzij. et al), *Gobi-Alt.* (Dundu-Saikhan mountains, south-west. slope of upper and middle belts, July 7, 1909—Czet.; Dzun-Saikhan mountains, in upper mountain belt, July 26; Dundu-Saikhan summit, on rocks, Aug. 19; same site, on dry river bed, Aug. 23; same site, on rocks in upper Yalo creek valley, Aug. 24—1931, Ik.-Gal.).

IIA. **Junggar:** *Tien Shan* (Naryngol, June 10, 1879—A. Reg.; 20 km south of Nyutsyuan'tsza, No. 245, July 18; Koisu river, in Ulansu region, No. 105, July 17; 36 km south-east of Nyutsyuan'tsza, No. 467, July 19; 20 km south of Dibukhe, No. 1776, Aug. 31—1957, Kuan; from Bortu to Khomote timber works, under spruce grove, No. 7033, Aug. 4, 1958—Lee and Chu).

IIIC. **Pamir** (Tagdumbash-Pamir, at confluence of Kara-Chukur and Ilyk-su rivers, on coastal rubble, July 16, 1901—Alekseenko; Kok-Muinak pass, on ascent along brook, rocks, July 27, 1909—Divn.).

General distribution: Jung.-Tarb., Nor. and Cent. Tien Shan; Arct. (Asian), West. Sib. (Altay), East. Sib., Far East, Nor. Hong., Himalayas (west.).

Note. In all probability, it hybridizes with *M. apetalum* but, in doubtful cases, always differs distinctly from the latter in seeds even when immature.

4. **M. glandulosum** (Maxim.) Will. in J. Linn. Soc. (London.) Bot. 34 (1899) 430; Hemsl. in J. Linn. Soc. (London) Bot. 36 (1904) 494; Pamp. Fl. Carac. (1930) 107; Hand.-Mazz. Symb. Sin. 7 (1929) 208; Rehder and Kobuski in J. Arn. Arb. 14 (1933) 9. —*Lychnis glandulosa* Maxim. Fl. Tangut. (1889) 83. —Ic.: Maxim. Fl. Tangut. tab. 29, fig. 1.

Described from Qinghai. Type in St.-Petersburg (LE). Plate 4, fig. 2.

Along banks of rivers and lakes.

IIIA. **Qinghai:** *Amdo* (around Khagomi in upper Huang He, 2100 m, on pebble bed, July 5, 1880—Przew., typus !; "Radja and Yellow River gorges, No. 14121—Rehder and Kobuski, l.c.)".

IIIB. **Tibet:** *Weitzan* (south. bank of Dzharin-nor lake, 4050 m, Aug. 5, 1884—Przew.).

General distribution: endemic.

5. **M. integripetalum** L.H. Zhou in Fl. Xizang. 1 (1983) 715. —Ic.: Fl. Xizang. 1, tab. 229, fig. 7-9.

Described from Tibet. Type in Beijing (PE).

On alpine meadows.

IIIB. Tibet: *South.* ("Lhasa, alt. 5000 m, 1908, Chang Young-tain and others", typus!—L.H. Zhou, l.c.).

General distribution: endemic.

M. multifurcatum (C.L. Tang) Yu. Kozhevn. comb. nov. —*Silene multifurcata* C.L. Tang in Acta Phytotax. Sin. 24, 5 (1986) 391. —*Melandrium fimbriatum* auct. non (Wall. ex Benth.) Walp.; L.H. Zhou in Fl. Xizang. 1 (1983) 719. —**Ic.:** Fl. Xizang. 1, fig. 231; C.L. Tang, l.c. fig. 3.

Described from Tibet. Type in Beijing (PE).

Note. Cited for south-east. Tibet (Xizang., Guona Xian, Aug. 26, 1975—Tang, l.c.) but probably occurs in south. Tibet.

6. **M. neocaespitosum** J.W. Tsui in L.H. Zhou in Fl. Xizang. 1 (1983) 720. Described from Tibet. Type in Beijing (PE).

On alpine meadows, up to 5000 m.

IIIB. Tibet: *Weitzan* ("An'do"—L.H. Zhou, l.c.), *South.* ("Gyangze, alt. 4750 m, Qinghai-Xizang Complex Exp. 2877, typus !; "Lhassa"—L.H. Zhou, l.c.)".

General distribution: endemic.

17. **Gypsophila** L.

Sp. pl. (1753) 406

1. Annual plants with long capilliform pedicels .... 2.
+ Perennial plants with relatively short or very short pedicels .... 3.
2. Plant with glandular pubescence. Calyx 3.5-4 mm long, with sharp teeth. Length of petals differs little from that of calyx or equal to it .... 5. **G. floribunda** (Kar. et Kir.) Turcz. ex Ledeb.
+ Plant pubescent with eglandular hairs, mainly in lower part. Calyx 2.5-3 mm long, with blunt teeth. Petals twice or almost twice longer than calyx .... 7. **G. muralis** L.
3. Cauline leaves large, ovoid or oblong-oval, up to 8 (9) cm long, 1.5 cm broad .... 4.
+ Cauline leaves usually small, lanceolate or linear, not more than 5 cm long, 1 cm broad .... 6.
4. Stems ascending. Leaves semiamplexicaul, pubescent or glabrous. Inflorescence paniculate, widespread, with glabrous branches; pedicels 3-6 times longer than calyx; bracts grassy ....
 .... 12. **G. trichotoma** Wend.
+ Stems erect. Leaves not amplexicaul at base, glabrous. Inflorescence corymbose-paniculate, not spreading; pedicels as long as calyx or not more than twice longer .... 5.

5. Bracts scarious. Inflorescence with dense-pubescent branches. Leaves generally more than 5 cm long .... 1. **G. altissima** L.
+ Bracts grassy, oval-lanceolate, sharp. Inflorescence branches glabrous. Leaves generally about 5 cm long... ....
.... 8. **G. oldhamiana** Miq.
6. Stems spreading, many, forming loose tufts .... 7.
+ Stems erect, not many, sometimes emerging from dense pulvinoid mat .... 8.
7. Leaves 6-15 mm long. Calyx 3.5-5 mm long. All surface parts of plant with dense-glandular pubescence......11. **G. sericea** (Ser.) Fenzl.
+ Leaves 4-6 mm long. Calyx 3-3.5 mm long. Stem and leaf margin pubescent with short eglandular hairs with small admixture of glandular .... 6. **G. microphylla** (Schrenk) Fenzl.
8. Flowers terminal and axillary. Stem forming bunches, not taller than 10 cm, emerging from strong rhizome. Entire plant densely pubescent with eglandular and glandular hairs ....
.... 4. **G. desertorum** (Bunge) Fenzl.
+ Inflorescence many-flowered, repeatedly branched or capitate. Bunches of stems emerging from stout rhizome, longer than 10 cm or plants forming mat. Pubescence only on some parts of plant or absent .... 9.
9. Plant with long branches in upper half, with extremely large number of small flowers. Calyx about 1.5 mm long; sepals blunt, broad-scarious along margin, with barely visible nerve. Lower stem dense-pubescent .... 9. **G. paniculata** L.
+ Upper half of plant poorly branched. Calyx more than 2 mm long; sepals sharp, narrow-scarious along margin, with prominent mid-nerve .... 10.
10. Inflorescence branches pubescent with glandular hairs. Radical leaves lanceolate, up to 1 cm broad ....
.... 3. **G. cephalotes** (Schrenk) Will.
+ Inflorescence branches glabrous. Radical leaves linear or lanceo-late-linear .... 11.
11. Plant pulvinoid, with 5-17 cm long peduncles. Inflorescence terminal, capitate, solitary or with 2-4 lateral axillary heads.......
.... 2. **G. capituliflora** Rupr.
+ Plant not pulvinoid. Peduncles up to 35 cm long. Inflorescence paniculate .... 10. **G. patrinii** Ser.

1. **G. altissima** L. Sp. pl. (1753) 407; Fenzl in Ledeb. Fl. Ross. 1 (1842) 289; Turcz. in Bull. Soc. natur. Moscou, 15, 3 (1842) 569; Will. in J. Bot. (London) 27 (1889) 325; Schischk. in Fl. SSSR, 6 (1936) 750; Opred. rast. Sr. Azii [Key to Plants of Mid. Asia] 2 (1971) 286. —*G. fastigiata* α *altissima*

Regel in Acta Horti Petrop. 5 (1877) 245; Maxim. Enum. pl. Mong. (1889) 87.

Described from West. Siberia. Type in London (Linn.).

In meadowy sections of mountain slopes.

IIA. Junggar: *Tien Shan* (Sairam, June 17, 1879; Dzhagastai, June 1881—A. Reg.). General distribution: Europe, Caucasus, West. Sib., East. Sib.

2. **G. capituliflora** Rupr. in Osten-Sacken et Rupr. Sertum tiansch. (1869) 40; Schischk. in Fl. SSSR, 6 (1936) 768; Barkhoudah, Rev. Gypsophila, Bolanthus... (1962) 82; Ikonnik. Opred. rast. Pamira [Key to Plants of Pamir] (1963) 110; Opred. rast. Sr. Azii [Key to Plants of Mid. Asia] 2 (1971) 288; Claves pl. Xinjiang. 2 (1983) 255. —*G. pamirica* Preobr. in Bull. Jard. bot. St.-Petersb. 16 (1916) 181. —*G. semiglobosa* Czerniak. in Not. syst. (Leningrad) 3 (1922) 129. —*G. dschungarica* Czerniak. ibid, 3 (1922) 130; Grub. Konsp. fl. MNR [Conspectus of Flora of Mongolian People's Republic] (1955) 133; Opred. rast. Sr. Azii [Key to Plants of Mid. Asia] 2 (1971) 289; Grub. Opred. rast. Mong. [Key to Plants of Mongolia] (1982) 105; Fl. Tadzh. 3 (1968) 600. —*G. acutifolia* β *gmelinii* sensu Maxim. non Regel, Fl. Tangut. (1889) 81; id. Enum. pl. Mong. (1889) 86. —Ic.: Fl. Tadzh. 3, Plate 92, fig. 2.

Described from East. Kazakhstan (Tien Shan). Type in St.-Petersburg (LE). Plate 10, fig. 4.

On rocky slopes.

IA. Mongolia: *Mong. Alt.* (in Dzusylyn gorge of Adzhi-Bogdo mountain range, June 15; north of Taishir-ol mountain range, July 2; in pass between Saksa and Tatal rivers, June 26—1877, Pot.; east. extremity of Gichigine-nuru mountain range, 3 km south-west of Amani-bulak spring on road from Bain-Undur somon, Aug. 27; 9 km south-west of Tsagan-olom on road to Yusun-Bulak, on limestone outcrop, Aug. 31—1948, Grub.; south. extremity of Arshantyn-nuru mountain, Bilut-ula mountains, July 20; Ulyastyin-gol river valley, lower and midcourse, July 8, 1984—Dariima, Kam.), *Gobi-Alt.* (west. extremity of Bain-Tsagan-ula 21 km from Bain-Tsagan somon on road to Delger somon, Aug. 28, 1948—Grub.), *Alash. Gobi* (along west. slope of central Alashan, on exposed mountain descents, July 9, 1873—Przew.).

IB. Kashgar: *Nor.* (Valle de Roe-koa sur Ak-cou, 1300-1800 m, July 28-30, 1900—Brocherel; Uch-Turfan, June 6, 1908—Divn.).

IIA. Junggar: *Tien Shan* (Sairam, July 1877—A. Reg.), *Jung. Gobi* (nor. Oshigini-usu, on smoothed granitic finely hummocky area, June 30, 1947—Yun.). General distribution: Tien Shan; Mid. Asia, China (Nor.-West.).

3. **G. cephalotes** (Schrenk) Will. in J. Bot. (London) 27 (1889) 323; Schischk. in Fl. SSSR, 6 (1936) 752; Grub. Konsp. fl. MNR [Conspectus of Flora of Mongolian People's Republic] (1955) 133; Ikonnik. Opred. rast. Pamira [Key to Plants of Pamir] (1963) 110; Opred. rast. Sr. Azii [Key to Plants of Mid. Asia] 2 (1971) 286; Podlech and Angers, in Mitt. Bot. Staatssaml. München, 13 (1977) 415; Grub. Opred. rast. Mong. [Key to Plants of Mongolia] (1982) 105. —*G. fastigiata* β *cephalotes* Schrenk in Fisch. et

Mey. Enum. pl. nov. (1841) 92; Fenzl in Ledeb. Fl. Ross. 1 (1842) 299; Kar. et Kir. in Bull. Soc. natur. Moscou, 15, 1 (1842) 165. —*G. transalaica* Ikonnik. in Novit. syst. pl. vasc. 12 (1975) 200. —Ic.: Fl. SSSR, 6, Plate 47, fig. 1.

Described from East. Kazakhstan (Jung. Ala Tau), Type in St.-Petersburg (LE). Plate 10, fig. 3.

On rocky slopes and alpine meadows.

IA. **Mongolia:** *Mong. Alt.* (midcourse of Bidzhi-gol river, valley slope, birch grove, near spring, Aug. 10, 1947—Yun.; Bulgan-gol river basin, Ulyastyin-gol gorge, nor. tributaries in upper courses, July 11, 1984—Da2riima, Kam.).

IB. **Kashgar:** *West.* (Sulu-Sokal valley, 25 km east of Irkeshtam, upper portion of juniper zone, 2800-2900 m, July 26, 1935—Olsuf'ev).

IIA. **Junggar:** *Cis-Alt.* (Qinhe, No. 1302, Aug. 2, 1956—Ching), *Tien Shan* (Talki, Aug.; Dzhagastai, Aug. 8, 1877; Agyaz river, June 26; Chapchal pass, June 28, 1878; Nilki brook in Kash river basin, June 8; Bogdo, July 24; on Chapchagai—Borgata road along nor. side of Kash river valley, July 6; Arystyn, July 15, 1879—A. Reg.; Urtas-Aksu, Kul'dzha region, June 17; near Sairam lake, July 23; Urten-muzart pass, Aug. 3, 1878—Fet.; mountains near Santash pass, on Tyubu river, June 20, 1893—Rob.; on Ven'tsyuan'—Syata road, No. 1560, Aug. 15; in Chzhaosu area from Chzhaosu on Teme, Aug. 11; east of Syata, No. 1340, Aug. 11; horse-breeding farm near Chzhaosu (=Kalmak-Kure), in intermontane valley, No. 3240, 808, Aug. 11; Ili—Chapchal, Dzhagastai, No. 3214, Aug. 8—1957, Kuan).

General distribution: Jung.-Tarb., Tien Shan.

Note. Some references to the find of *G. acutifolia* Fisch. ex Spreng. (Limpr. in Feddes Repert. 12 (1922) 364; Fl. desert. Sin. 1 (1985) 473; tab. 172, fig. 5-7; Rast. pokr. Vnutr. Mong. [Vegetational Cover of Inner Mongolia] (1985) 73), most probably are based on incorrect identification of *G. cephalotes*. At the same time, we are not certain if the latter species is simply not a race of former, characterized by compact inflorescence.

4. **G. desertorum** (Bge.) Fenzl in Ledeb. Fl. Ross. 1 (1842) 292; Maxim. Enum. pl. Mong. (1889) 85; Kryl. Fl. Zap. Sib. 5 (1931) 1087; Schischk. in Fl. SSSR, 6 (1936) 740; Grub. Konsp. fl. MNR [Conspectus of Flora of Mongolian People's Republic] (1955) 133; Ma Yu-chuan in Fl. Intramong. 2 (1978) 186; Grub. Opred. rast. Mong. [Key to Plants of Mongolia] (1982) 105; Fl. desert. Sin. 1 (1985) 469. —*Heterochroa desertorum* Bge., Suppl. Alt. (1836) 29. —Ic.: Fl. SSSR, 6. Plate 46, fig. 4; Fl. Intramong. 2, tab. 99; Fl. desert. Sin. 1, tab. 172, fig. 8-9; Rast. pokr. Vnutr. Mong. [Vegetational Cover of Inner Mongolia] (1985) 74.

Described from Altay. Type in St.-Petersburg (LE). Map 7.

On arid and barren steppes, sand, rubbly slopes, pebble beds, among rocks.

IA. **Mongolia:** *Khobd., Mong. Alt., Cent. Khalkha, Depr. Lakes, Valley Lakes, Gobi-Alt., East. Gobi, Alash Gobi.*

General distribution: West. Sib. (Altay), East. Sib. (Tuva), Nor. Mong. (Fore Hubs., Hent., Hang., Mong.-Daur.), China (Nor.-West.).

5. **G. floribunda** (Kar. et Kir.) Turcz. ex Ledeb. Fl. Ross. 1 (1842) 755; Boiss. Fl. or. 1 (1867) 553; Kryl. Fl. Zap. Sib. 5 (1931) 1086; Schischk. in Fl. SSSR, 6 (1936) 775; Opred. rast. Sr. Azii [Key to Plants of Mid. Asia] 2 (1971) 290; Kamelin et al in Byull. Mosk. obshch. isp. prir., otd. biol. 90, 5 (1985) 115. —*Dichoglottis floribunda* Kar. et Kir. in Bull. Soc. natur. Moscou, 15, 1 (1842) 165. —*Saponaria floribunda* (Kar. at Kir.) Boiss. Diagn. pl. or., nov. ser. 2, 1 (1853) 70; Fl. Tadzh. 3 (1968) 639.

Described from Kazakhstan. Type in Moscow (MW).

On rocky slopes, gorges.

IIA. Junggar: *Jung. Gobi* (right bank of Bulgan-gol river, below Bulgan somon, Dzun-Khara-ula low mountain, in barren low mountains, July 18, 1984—Dariima, Kam.).

General distribution: Fore Balkh., Nor. Tien Shan; West. Sib. (Altay), Mid. Asia (Pam.-Alay).

6. **G. microphylla** (Schrenk) Fenzl in Ledeb. Fl. Ross. 1 (1842) 291; Schischk. in Fl. SSSR, 6 (1936) 740; Opred. rast. Sr. Azii [Key to Plants of Mid. Asia] 2 (1971) 284; Claves pl. Xinjiang. 2 (1983) 255. —*Heterochroa microphylla* Schrenk, Enum. pl. soong. (1841) 92; Kar. et Kir. in Bull. Soc. natur. Moscou, 15, 1 (1842) 165. —**Ic.:** Fl. SSSR, 6, Plate 46, fig. 5.

Described from East. Kazakhstan (Jung. Ala Tau). Type in St.-Petersburg (LE). Plate 10, fig. 1.

On rubbly slopes and talus.

IIA. Junggar: *Tien Shan* (Sairam lake basin, July 12, 1878—Fet.), *Jung. Alat.* (Borotala river basin, below Koketau pass, July 21, 1909—Lipsky).

General distribution: Jung.-Tarb., Nor. Tien Shan.

7. **G. muralis** L. Sp. pl. (1753) 408; Fenzl in Ledeb. Fl. Ross. 1 (1842) 288; Kar. et Kir. in Bull. Soc. natur. Moscou, 15, 1 (1842) 165; Maxim. Enum. pl. Mong. (1889) 85; Kryl. Fl. Zap. Sib. 5 (1931) 1085; Schischk. in Fl. SSSR, 6 (1936) 774. —*G. stepposa* Klok. in Zhurn. Russk. bot. obshch. 6 (1921) 137. —*Psammophila muralis* (L.) Fourr. in Ann. Soc. Linn. Lyon, 16 (1868) 345. —*Psammophiliella muralis* (L.) Ikonnik. in Novit. syst. pl. vasc. 13 (1976) 116.

Described from Scandinavia. Type in London (Linn.).

On garbage, solonetzes.

IIA. Junggar: *Zaisan* (on Chern. Irtysh river, 1876—Pot.).

General distribution: Fore Balkh.; Mediterr., Balk.-Asia Minor, Mid. Asia, West. Sib., East. Sib., Far East.

8. **G. oldhamiana** Miq. in Ann. Mus. Bot. Lugd.-Batav. 3 (1867) 187; Forbes et Hemsl. Index Fl. Sin. (1888) 64; Claves pl. Xinjiang. 2 (1983) 255.

Described from Japan. Type in Utrecht (U).

In steppes.

IIA. Junggar: *Tarb.* ("Dachen"—Claves, l.c.), *Tien. Shan* ("Chzhao-su, Sin'yuan'", "Gunlyu"—Claves, l.c.).

General distribution: China (Nor.), Japan.

Note. We have not studied any material of this species and we are not certain if it is distinct enough from *G. altissima.*

9. **G. paniculata** L. Sp. pl. (1753) 407; Fenzl in Ledeb. Fl. Ross. 1 (1842) 297; Maxim. Enum. pl. Mong. (1889) 86; Kryl. Fl. Zap. Sib. 5 (1931) 1091; Schischk. in Fl. SSSR, 6 (1936) 749; Grub. Konsp. fl. MNR [Conspectus of Flora of Mongolian People's Republic] (1955) 133; Opred. rast. Sr. Azii [Key to Plants of Mid. Asia] 2 (1971) 285; Grub. Opred. rast. Mong. [Key to Plants of Mongolia] (1982) 105; Claves pl. Xinjiang. 2 (1983) 255; Fl. desert. Sin. 1 (1985) 471. —**Ic.:** Fl. SSSR, 6, Plate 47, fig. 3; Fl. desert. Sin. 1, tab. 172, fig. 10-11.

Described from Siberia. Type in London. (Linn.).

In sand steppes and on poorly fixed sand.

IA. Mongolia: *Mong. Alt.* (bank of Bulugun river above Dzhirgalant river estuary, July 25, 1898—Klem.), *Depr. Lakes* (west. fringe of Ubsa lake basin, Khara-bura sand, Oct. 4, 1931—Bar.; Borig-del' sand south-east of Bain-nur lake, July 25, 1945—Yun.; Borig-del' sand, Tsagan-del' area 22 km west of Dzun-Gobi somon, Aug. 21, 1971—Grub., Ulzij. et al; central portion of Borig-del' sand massif, July 13; 58 km east of Ulangom, July 22—1973, Banzragch et al).

IIA. Junggar: *Cis-Alt.* (vicinity of Chingil' town, No. 1621, Aug. 8; around Koktogai town, No. 2102, Aug. 18—1956, Ching), *Tarb.* (between Lasta and Tumanda rivers, Aug., 7; between Tumanda and Semiz-chiem, Aug. 8, 1876—Pot.; 15 km nor. of Dachen town, No. 1493, Aug. 12, 1957—Kuan), *Jung. Gobi* (Kran river lower course, steppe, June 26, 1908—Sap.; on Khutubi—Urumchi road, in Gobi, No. 5056, Sept. 21, 1957—Kuan), *Zaisan* (Alkabek, sand, Aug. 6, 1908—Fedchenko; Chernyi Irtysh river, left bank, Dzhelkaidar area, sand, June 8; same site, Maikain area, hummocky sand, June 7, 1914—Schischkin; 9 km nor.-west of Araak, on fixed sand, July 10, 1959—Lee and Chu).

General distribution: Aralo-Casp., Fore Balkh.; Europe, Caucasus, West. Sib., China (Nor.-West.), Nor. Amer. (introduced).

Note. In plants of this species, petals could be white and violet.

10. **G. patrinii** Ser. in DC. Prodr. 1 (1824) 353; Kryl. Fl. Zap. Sib. 5 (1931) 1089; Schischk. in Fl. SSSR, 6 (1936) 767; Hao in Bot. Jahrb. 68 (1938) 595; Grub. Konsp. fl. MNR [Conspectus of Flora of Mongolian People's Republic] (1955) 134; Opred. rast. Sr. Azii [Key to Plants of Mid. Asia] 2 (1971) 288; Grub. Opred. rast. Mong. [Key to Plants of Mongolia] (1982) 105; Claves pl. Xinjiang. 2 (1983) 256; Fl. desert. Sin. 1 (1985) 471. —*G. davurica* Turcz. ex Fenzl in Ledeb. Fl. Ross. 1 (1842) 294; Schischk. in Fl. SSSR, 6 (1936) 768; Walker in Contribs. U.S. Nat. Herb. 28, 4 (1941) 613; Grub. Konsp. fl. MNR [Conspectus of Flora of Mongolian People's Republic] (1955) 133; Ma Yu-chuan in Fl. Intramong. 1 (1978) 186, s. str. et var. *angustifolia* Fenzl; Grub. Opred. rast. Mong. [Key to Plants of Mongolia] (1982) 105. —*G. patrinii* subsp. *davurica* (Fenzl) Yu. Kozhevn. in Novit. syst. pl. vasc. 22 (1985) 111. —*G. gmelinii* δ *davurica* Turcz. in Bull. Soc. natur. Moscou, 15, 3 (1842) 571;

Maxim. Enum. pl. Mong. (1889) 86; Fl. desert. Sin. 1 (1985) 471. —*G. gmelinii* Bge. α *patrinii* and β *caespitosa* in Ledeb. Fl. Alt. 2 (1830) 128; Kar. et Kir. in Bull. Soc. natur. Moscou, 15, 1 (1842) 165; Turcz. in Bull. Soc. natur. Moscou, 15, 3 (1842) 570; Walker in Contribs. U.S. Nat. Herb. 28, 4 (1941). —Ic.: Fl. desert. Sin. 1, tab. 171, fig. 1-2, tab. 172, fig. 1-4.

Described from Altay. Type in Geneva (G).

On riverine pebble beds, steppe and barren rocky and rubbly slopes, on sand, rocks, meadows.

IA. **Mongolia:** *Khobd., Mong. Alt., Cent. Khalkha, East. Mong., Depr. Lakes, Val. Lakes, Gobi-Alt., Alash. Gobi, Ordos.*

IIA. **Junggar:** *Jung. Alat.* (Dzhair mountain range, 1-1.5 km north of Otu settlement, on granite conical hillocks, Aug. 4, 1957—Yun. et al; on Ebi-nur—Toli road, No. 1765, Aug. 17; in Toli region, on slope, No. 2473, Aug. 4, 1957—Kuan), *Tien Shan, Jung. Gobi* (vicinity of Urumchi town, Khunmiodza upland, Sept. 10, 1929—Pop.; south of Barbagai, No. 2856, Sept. 8, 1956—Ching; 40 km south of Chingil', No. 4091, Sept. 2; Barbagai, No. 2803, 2786, Sept. 4; from Barbagai to Burchum, No. 2803, Sept. 4, 1956—Ching; Tsitai—Beidashan', No. 15242, Sept. 29, 1957—Kuan; on Ertai—Tsitai road, Aug. 3, 1959—Lee and Chu; south. extremity of Khuvchiin-nuru mountain, Aug. 1, 1984—Dariima, Kam.), *Zaisan* (on Chern. Irtysh, Aug. 28, 1876; on rocks in Yamadzhin mountains, July 11, 1877—Pot.; Chern. Irtysh river, right bank below Burchum river, Sary-Dzhasyk-Kiikpai well, June 15; same site, left bank of Dzhelkaidar, June 9, 1914—Schischk.).

IIIA. **Qinghai:** *Amdo* (on upper Huang He, on Mudzhik-khe river, 2700-2850 m alt., June 30, 1880—Przew.).

General distribution: Jung.-Tarb., Nor. Tien Shan; Europe, Mid. Asia, West. Sib., East. Sib., Far East, Nor. Mong., China (Dunbei, Nor., Nor.-West., South-West.).

Note. This species exhibits considerable polymorphism, especially in its life form.

11. **G. sericea** (Ser.) Fenzl in Endl. Gen. pl. 1 (1836-1840) 972; Kryl. Fl. Zap. Sib. 5 (1931) 1087; Schischk. in Fl. SSSR, 6 (1936) 739. —*Arenaria sericea* Ser. in DC. Prodr. 1 (1824) 414. —*Heterochroa petraea* Bge. in Ledeb. Fl. Alt. 2 (1830) 132. —*Gypsophila petraea* Fenzl in Ledeb. Fl. Ross. 1 (1842) 291, non Reichb.; Maxim. Enum. pl. Mong. (1889) 85.

Described from Siberia. Type in Geneva (G).

On rocky and rubbly slopes; on rocks, in sparse forests.

IIA. **Junggar:** *Cis-Alt.* (on rocks in Kandagatai gorge, Sept. 14, 1876—Pot.; mountain slope near Dzurkhe river, one of the summits of Tsagan-gol, sparse larch forest, July 30, 1898—Klem.; Kungeity river valley, right tributary of Kara-Irtysh river, in forest glades, July 8, 1908—Sap.; in Shara-Sume region, 2100 m, No. 1178, Aug. 2; Qinhe—Chzhunkhaitsze and Qinhe, in steppe, No. 1303, 1264, Aug. 3, 1956—Ching; 30 km north of Koktogai, right bank of Kairta river, Kuidyn river valley, forest belt, July 15, 1959—Yun. and I.-F. Yuan').

General distribution: West. Sib. (Altay), East. Sib.

12. **G. trichotoma** Wend. in Linnaea, 11 (1837) 92; Fenzl in Ledeb. Fl. Ross. 1 (1842) 297; Maxim. Enum. pl. Mong. (1889) 87; Kryl. Fl. Zap. Sib. 5

(1931) 1090; Schischk. in Fl. SSSR, 6 (1936) 759; Grub. Konsp. fl. MNR [Conspectus of Flora of Mongolian People's Republic] (1955) 134; id. Opred. rast. Mong. [Key to Plants of Mongolia] (1982) 104; Claves pl. Xinjiang. 2 (1983) 256; Fl. desert. Sin. 1 (1985) 469. —Ic.: Fl. Kazakhst. 3, Plate 42, fig. 4; Fl. desert. Sin. 1, tab. 171, fig. 3-5.

Described from Mid, Asia. Type in London (?). Plate 10, fig. 2.

On solonetzic sand, coastal pebble beds.

IA. Mongolia: *Depr. Lakes* (around Ubsa lake, Sept. 22, 1879—Pot.; Ubsa lake, sand bank—Klem.; on nor. bank of Ubsa lake, barren steppe, July 3, 1892—Kryl.).

IIA. Junggar: *Cis-Alt.* (vicinity of Chingil' (Qinhe) town, Aug. 8; in Shara-Sume region, No. 2302, Aug. 20, 1956—Ching.), *Tarb.* (on Tumandy—Semiz-chii road, Aug. 8, 1876—Pot.; south of Dachen town, on bank of brook in desert, No. 2872, Aug. 11, 1957—Kuan), *Jung. Alat.* (left bank of lower Borotala, Aug. 22; Uch-Tyube, Aug. 1878— A. Reg.), *Tien Shan* (upper Ili river, Aug. 29, 1876; near Kash river, July 12, 1877— Przew.; arid steppe, 3600 m, June 12, 1893—Rob.; 2 km south of Sykeshu, No. 4809, Sept. 1, 1957—Kuan), *Jung. Gobi* (Barbagai, No. 2807, Sept. 11, 1956—Ching; Datsyuan'-gou village in Savan area, on roadside, No. 1214, July 4, 1957—Kuan), *Zaisan* (Chern. Irtysh river, right bank below Burchum river, Sary-dzhasyk-kol, Kiikpai, willow groves, June 15; same site, left bank, Maikain area, hummocky sand, June 7, 1914—Schischk.), *Dzhark.* (Kul'dzha, July 6; Suidun, July 16; west of Kul'dzha, July 1877—A. Reg.; Kul'dzha town, Ili river valley near pass, on pebble-beds, Aug. 21, 1957—Yun. and I.-F. Yuan').

General distribution: Aralo-Casp., Fore Balkh., Jung.-Tarb., Nor. Tien Shan; Europe, Fore Asia, Caucasus, West. Sib.

Note. Two varieties are distinguished: var. *glabra* Fenzl with glabrous stems and var. *pubescens* Fenzl with stems pubescent in lower part.

18. **Petrorhagia** (Ser.) Link

Handb. 2 (1831) 235. —*Gypsophila* sect. *Petrorhagia* Ser. in DC. Prodr. 1 (1824) 354.

P. alpina (Habl.) P.W. Ball et Heywood in Bull. Brit. Mus. (Bot.) 3 (1964) 145; P.W. Ball in Fl. Eur. 1 (1964) 187. —*Gypsophila alpina* Habl. Neues Nord. Beitr. 4 (1783) 57. —*G. stricta* Bge. in Ledeb. Fl. Alt. 2 (1830) 129. — *Tunica stricta* (Bge.) Fisch. et Mey. Ind. Sem. Horti Petrop. 4 (1837) 50; Kar. et Kir. in Bull. Soc. natur. Moscou, 15, 1 (1842) 165; Kryl. Fl. Zap. Sib. 5 (1931) 1094; Gorschk. in Fl. SSSR, 6 (1936) 778; Opred. rast. Sr. Azii [Key to Plants of Mid. Asia] 2 (1971) 293; Claves pl. Xinjiang. 2 (1983) 249. —*Dianthus recticaulis* Fenzl in Ledeb. Fl. Ross. 1 (1842) 287; Trautv. in Bull. Soc. natur. Moscou, 32, 1 (1860) 143. —Ic.: Fl. SSSR, 6, Plate 55. fig. 5.

Described from West. Siberia. Type in St.-Petersburg (LE).

On rubbly steppe slopes, sandy-pebbly banks of streams.

IB. Kashgar: *East.* (on Kucha-Urumchi road, on slope, No. 2574, Aug. 28, 1956— Ching).

IIA. Junggar: *Cis-Alt.* (Kurtu river, July 24, 1903—Gr.-Grzh.; 20 km nor.-west of Shara-Sume, midmountain belt, July 7, 1959—Yun. and I.-F. Yuan'; Qinhe, No. 1306,

Aug. 3, 1956—Ching), *Tarb.* (Kobuk river valley, larch forest, on brook, July 20, 1914—
Sap.; Dachen town, No. 1666, Aug. 14, 1957—Kuan), *Tien. Shan* (half-way between Ili
and Chapchal, Dzhagastai, No. 3184, Aug. 8, 1958–Lee and Chu; Dzhagastai, Aug. 7,
1877—A. Reg.; Bogdo mountain, July 24, 1878—A. Reg.; mountains near Santash and
Tyube river crossing, June 20, 1893—Rob.).

General distribution: Jung.-Tarb., Tien Shan; Balk.-Asia Minor, Fore Asia, Caucasus,
Mid. Asia, West. Sib. (Altay), East. Sib.

19. **Acanthophyllum** C.A. Mey.

Verzeichn. Pfl. Cauc. (1831) 210

A. pungens (Bge.) Boiss. Fl. Or. 1 (1867) 561; Kryl. Fl. Zap. Sib. 5 (1931)
1095; Schischk. in Fl. SSSR, 6 (1936) 784; Grub. Konsp. fl. MNR [Conspectus
of Flora of Mongolian People's Republic] (1955) 134; Opred. rast. Sr. Azii
[Key to Plants of Mid. Asia] 2 (1971) 295; Grub. Opred. rast. Mong. [Key to
Plants of Mongolia] (1982) 105; Claves pl. Xinjiang. 21 (1983) 250; Fl. desert.
Sin. 1 (1985) 468. —*A. spinosum* C.A. Mey. l.c. 210; Kar. et Kir. in Bull. Soc.
natur. Moscou, 15, 1 (1842) 166; Trautv. in Bull. Soc. natur. Moscou, 32, 1
(1860) 156; Maxim. Enum. pl. Mong. (1889) 94. —*Saponaria pungens* Bge. in
Ledeb. Fl. Alt. 2 (1830) 133. —**Ic.:** Fl. SSSR, 6, Plate 49, fig. 3; Grub. Opred.
rast. Mong. [Key to Plants of Mongolia] Plate 46, fig. 216; Fl. desert. Sin. 1,
tab. 169, fig. 4-5.

Described from West. Siberia (Irtysh river). Type in St.-Petersburg (LE).
Plate 5, fig. 3.

On rubbly slopes and sand.

IA. Mongolia: *Mong. Alt.* (Iter ad Chobdo, June 19, 1870—Kalning; trail of low
mountains on right bank of Bulgan-gol river, July 18, 1984—Dariima, Kam.).

IIA. Junggar: *Tarb.* (south. trail of Urkashar mountain range on Khobuk-Chuguchak
road, saxaul hammada, along gorges, Aug. 5, 1957—Yun. and I.-F. Yuan'), *Jung. Alat.*
(vicinity of Dzhair mountain range, 25-27 km north-east of Toli settlement on road to
Temirtam, trails of Tamgary-terek minor mountain range, desert, Aug. 5, 1957—Yun.
and I.-F. Yuan'; in Toli region, No. 887, Aug. 3, 1957—Kuan), *Tien Shan* (Aktyube north
of Kul'dzha, May 13, 1977; Khanakhai brook south-west of Kul'dzha, June 15, 1878—
A. Reg.; 35 km east of bridge on Kash river toward Sin'yuan' town, on roadsides and
on slope, No. 1106, Aug. 21, 1957—Kuan), *Jung. Gobi* (on Baitak-Bogdo mountains,
before Kyup spring, Aug. 8, 1898—Klem.; Tien-Shan-laoba, Myao-ergou, No. 2342,
Aug. 3; Dashitou, near Guchen, in desert, No. 2469, Oct. 1, 1957—Kuan), *Zaisan*
(Chern. Irtysh valley, coastal sand, June 6, 1903—Gr.-Grzh.; right bank of Cher. Irtysh
above Kran river estuary, June 27, 1908—Sap.).

General distribution: Fore Balkh., Jung.-Tarb., Tien Shan.

20. **Vaccaria** N.M. Wolf

Gen. Pl. Vocab. Char. Def. (1776) 111

V. hispanica (Mill.) Rauschert in Feddes Repert. 73, 1 (1966) 52.
—*V. segetalis* (Neck.) Garcke in Ascers. Fl. Prov. Brand. 1 (1864) 84; Kryl. Fl.

Zap. Sib. 5 (1931) 1096; Gorschk. in Fl. SSSR, 6 (1936) 802; Grub. Konsp. fl.
MNR [Conspectus of Flora of Mongolian People's Republic] (1955) 134;
Ma Yu-chuan in Fl. Intramong. 2 (1978) 193; Grub. Opred. rast. Mong. [Key
to Plants of Mongolia] (1982) 105; Zhou Li-hou in Fl. Xizang. 1 (1983) 741;
Claves pl. Xinjiang. 2 (1983) 264; Fl. desert. Sin. 1 (1985) 463. —*V. pyramidata*
Medic. Phil. bot. 1 (1789) 96; Ikonnik. Opred. rast. Pamira [Key to Plants of
Pamir] (1963) 111; Opred. rast. Sr. Azii [Key to Plants of Mid. Asia] 2 (1971)
300; Podlech and Angers, in Mitt. Bot. Staatssaml. Munchen, 13 (1977) 418.
—*Saponaria hispanica* Mill. Gard. Dict. ed. 8 (1768) in erratis. —*S. segetalis*
Neck. Dilec. Gallo-Belg. 1 (1768) 194. nom. illeg. —*S. vaccaria* L. Sp. pl.
(1753) 409; Forbes et Hemsl. Index Fl. Sin. (1886) 64; Maxim. Enum. pl.
Mong. (1889) 87. —**Ic.**: Fl. SSSR, 6, Plate 55, fig. 8; Grub. Opred. rast. Mong.
[Key to Plants of Mongolia] Plate 46, fig. 215; Fl. Intramong. 2, tab. 102, fig.
1-3; Fl. desert. Sin. 1, tab. 169, fig. 1-3.

Described from Europe. Type in London (Linn.).

In fields and kitchen gardens, roadsides, garbage.

IA. Mongolia: *Mong. Alt.* (near Kobdo, weed in kitchen gardens, Aug. 11, 1898;
around Tsakhir-bulak spring, July 18, 1894—Klem.; Bodkhon-gol area, Aug. 11, 1945;
Bulugun river valley near winter camp of Bulugun somon, Aug. 1949—Yun.), *Depr.
Lakes* (Shargain-Gobi desert, Khalyun-gol river, on fallow land, Aug. 16, 1930—Pob.;
Kharkira river valley, 15 km north of Chianai, Aug. 22; alongside road 10 km before
Ulan-Tyube north-west of Ulangom, Aug. 12, 1931—Shukhardin), *Gobi-Alt.* (Leg lowland
near Bain-Leg somon camp, in hedges and along canal, Aug. 25, 1948—Grub.;
submontane plain on south, slope of Dzun-Saikhan mountain range 45 km nor.-east of
Bain-Dalai somon, Sept. 15, 1951—Kal.), *Alash. Gobi* (Dyn-yuan'in oasis, June 5, 1908—
Czet.; vicinity of In'chuan' town, 45 km south of town, June 3, 1957—Petr.).

IB. Kashgar: *Nor.* (Uch-Turfan, in hedge, June 9, 1908—Divn.), *South.* (Kargalyk
oasis, in plantations, July 8, 1889—Rob.).

IIA. Junggar: *Cis-Alt.* (20 km nor.-west of Shara-Sume, July 7, 1959—Yun. and I.-F.
Yuan'; Qinhe, No. 1640, Aug. 11, 1956—Ching), *Dzhark.* (Ili bank west of Kul'dzha,
July; Kul'dzha, July 5, 1877; between Suidun and Ili river, June 2, 1878—A. Reg.;
vicinity of Kul'dzha, in grass, June 3, 1878—Larionov), *Jung. Gobi* (Bulugun river lowland,
plantations among *Colocasia*-winterfat barren land, July 28, 1947—Yun.; right bank of
Manas river, 3-4 km south of Mo-Savan area on; road to Da-Myao, on fallow land, July
10, 1957—Yun. et al; 7 km south of Manas, midcourse of Dasi river, along field boundary,
No. 1343, July 12, 1957—Kuan; east of Barbagai, in Kran river floodplain, July 8,
1959—Lee and Chu), *Zaisan* (between Kara area and village on Kabe river, steppe, June
16, 1914—Schischk.).

General distribution: Aralo-Casp., Fore Balkh., Nor. and Cent. Tien Shan; Europe,
Mediterr., Balk.-Asia Minor, Fore Asia, Caucasus, Mid. Asia, West. Sib., Far East, Nor.
Mong., China (Nor., Nor.-West., Cent.), Himalayas, Korean peninsula, Japan, Indo-
Mal., Nor. Amer. (introduced), N. Zealand (introduced).

21. Dianthus L.

Sp. pl. (1753) 409

1. Petals laciniated into many narrow lobes .... 2.
+ Petals entire, dentate on upper margin .... 4.

2.　Leaves acicular, spiny. Petals laciniated for 1/3–1/2 of their length. Glumes cover 1/4 of calyx, with short cusp .... 1. **D. arenarius** L.
+　Leaves not acicular, not spiny. Petals laciniated almost down to base. Glumes cover 1/3–1/2 of calyx, with long cusp, sometimes reaching tip of calyx .... 3.
3.　Stems solitary or few, grassy, emerging from creeping funiform rhizome. Leaves lanceolate or linear-lanceolate, up to 7 cm long. Calyx about 4 mm broad, rarely up to 5 mm, with relatively short oval-deltoid teeth ....5. **D. superbus** L.
+　Stems usually many, forming loose tuft, woody in lower part, emerging from stout rhizome. Leaves linear, often filiform, up to 3.5-4 cm long. Calyx about 3 mm broad, with narrow-deltoid teeth ....3. **D. crinitus** Smith.
4.　Annual plants with 6–15 mm broad lanceolate leaves, ciliate or glabrous along margin. Anthers dark-blue. Petals pink above, yellowish beneath .... 2. **D. chinensis** L.
+　Perennial plants with short-pubescent leaves not more than 6 mm broad. Anthers pink or red. Petals purple or white above, greenish beneath .... 5.
5.　Stems more than 1 mm thick, forming 25–40 cm tall loose tuft. Calyx (10) 12–23 (25) mm long, 4–5 mm broad. Petals projecting from calyx for 8–15 mm. Leaves 1.5–2.5 (3) mm broad ....
.... 6. **D. versicolor** Fisch. ex Link.
+　Stems less than 1 mm thick, forming 20-30 cm tall dense tuft. Calyx 12–14 mm long and about 3 mm broad. Petals projecting from calyx for 5–7 mm. Leaves not more than 1.5 mm broad ......
.... 4. **D. ramosissimus** Pall. ex Poir.

　1. **D. arenarius** L. Sp. pl. (1753) 413; Ledeb. Fl. Ross. 1 (1842) 284; Schischk. in Fl. SSSR, 6 (1936) 848. —*D. acicularis* Fisch. ex Ledeb. Fl. Ross. 1 (1842) 284; Kryl. Fl. Zap. Sib. 5 (1931) 1099; Opred. rast. Sr. Azii [Key to Plants of Mid. Asia] 2 (1971) 305; Claves pl. Xinjiang. 2 (1983) 252; Fl. desert. Sin. 1 (1985) 468.

Described from Europe. Type in London (Linn.).

On sand, rocky slopes and rocks.

IIA. Junggar: *Tien Shan* (Tsitai, Inin"—Claves pl, Xinjiang. l.c.). **General distribution:** East. Europe, West. Sib.

　Note. *D. acicularis*, "in fact a subspecies of *D. arenarius*, has been reported in our territory."

　2. **D. chinensis** L. Sp. pl. (1753) 411; Rohrb. in Linnaea, 36 (1870) 670; Forbes et Hemsl. Index Fl. Sin. 1 (1886) 63; Schischk. in Fl. SSSR, 6 (1936) 823; Hao in Bot. Jahrb. 68 (1938) 593; Walker in Contribs. U.S. Nat. Herb.

28, 4 (1941) 613; Claves pl. Xinjiang. 2 (1983) 252; Fl. desert. Sin. 1 (1985) 465, fig. 6-8.

Described from China. Type in London (Linn.).

Habitat type not known.

IIA. Junggar: *Tien Shan* (Kash river, Aug. 4, 1880—A. Reg., plants grown in 1882 in Pomologichesk garden in St.-Petersburg from seeds collected by A. Regel).

General distribution: China, Japan.

Note. The main characteristic of this species is its being an annual plant which fact emerges only from the reference of Linnaeus as the type is represented only by the upper parts of plants. Many taxonomists, including Maximowicz, ignored this reference and, as a result, there is no clear interpretation of this species so far.

3. **D. crinitus** Smith in Trans. Linn. Soc. Bot. 2 (1794) 300; Ledeb. Fl. Ross. 1 (1842) 283; Kar. et Kir. in Bull. Soc. natur. Moscou, 15, 1 (1842) 166; Trautv. in Bull. Soc. natur. Moscou, 32, 1 (1860) 143; Edgew. et Hook. f. in Fl. Brit. India, 1 (1875) 215; Henders. and Hume, Lahore to Yarkand (1873) 311; Maxim. Enum. pl. Mong. (1889) 84; Kryl. Fl. Zap. Sib. 5 (1931) 110; Schischk. in Fl. SSSR, 6 (1936) 850. —*D. soongoricus* Schischk. in Fl. URSS, 6 (1936) 899; Grub. Opred. rast. Mong. [Key to Plants of Mongolia] (1982) 105; Claves pl. Xinjiang. 2 (1983) 253; Fl. desert. Sin. 1 (1985) 466. —*D. kuschakewiczii* Rgl. et Schmalh. in Acta Horti Petrop. 5 (1887) 244; Will. in J. Linn. Soc. (London) Bot. 29 (1893) 397; Schischk. in Fl. SSSR, 6 (1936) 851; Fl. Kazakhst. 3 (1960) 427; Opred. rast. Sr. Azii [Key to Plants of Mid. Asia] 2 (1971) 305; Claves pl. Xinjiang. 2 (1983) 253. —Ic.: Fl. SSSR, 6, Plate 54, fig. 5; Grub. Opred. rast. Mong. [Key to Plants of Mongolia] Plate 46, fig. 217; Fl. desert. Sin. 1, tab. 170, fig. 1-2.

Described from Armenia. Type in Geneva (G). Plate 10, fig. 6.

On rocky and rubbly slopes, alpine steppes, rocks.

IIA. Junggar: *Jung. Alat.* (ascent to Kuzyun' pass, Aug. 2, 1908—Fedczenko; Dzhair mountain range, 5 km south-west of Boguta river on Otu-Aktama road, Aug. 3, 1957—Yun. et al; "Chzhaosu"—Claves pl. Xinjiang. l.c.), *Tien Shan* (Arasan bridge on Khanakhai river, June 24; near Kash river, Aug. 5, 1878—A. Reg.; Urtak-Sary west of Sairam lake, July 19; near Sairam lake, July 19, 1878—Fet.; upper Tekes river, July 25, 1893—Rob.; Sumbe river midcourse, July 9, 1912—Sap. and Schischk.), *Jung. Gobi* (3 km east of Yants-zykhai (Bain-gol) bridge on Manas-Shikho road, on gorge, July 7, 1957; 68 km from Ertai on Koktogai road, July 14, 1959—Yun. and I.-F. Yuan'; Bodonchin-gol river valley, Khara-Togo-khuduk well, about 1400 m, June 30, 1973— Golubkova and Tsogt; Ulyastain-gol, near east. bend at its exit in Bulgan-gol valley, on pebble bed, Aug. 16, 1979—Grub. et al), *Zaisan* (near Dyurbel'dzhin on Chern. Irtysh, Aug. 24, 1876—Pot.; Urta-Ulasty river, nor. foothill of Saur, June 18, 1908—Sap.; vicinity of Kenderlyk village, June 2; between Burchum river and Kaba settlement, Kiikpai well—Karoi area, June 15; between Karoi area and village on Kaba river, June 16, 1914—Schischk.), *Dzhark.* (Suidun, July 16, 1877—A. Reg.).

General distribution: Tien Shan; Fore Asia, Mediterr. (Cyprus), Caucasus.

Note. Forms intermediate to *D. crinitus* and *D. superbus*, probably of hybrid origin, occur.

4. **D. ramosissimus** Pall. ex Poir. Encycl. Suppl. 6 (1816) 130; Kryl. Fl. Zap. Sib. 5 (1931) 1102; Schischk. in Fl. SSSR, 6 (1936) 835; Grub. Konsp. fl. MNR [Conspectus of Flora of Mongolian People's Republic] (1955) 134; Opred. rast. Sr. Azii [Key to Plants of Mid. Asia] 2 (1971) 304; Grub. Opred. rast. Mong. [Key to Plants of Mongolia] (1982) 106. —*D. campestris* var. *glabra* Trautv. in Bull. Soc. natur. Moscou, 32, 1 (1860) 142; Maxim. Enum. pl. Mong. (1889) 84. —Ic.: Grub. Opred. rast. Mong. [Key to Plants of Mongolia] Plate 46, fig. 214.

Described from Siberia. Type in St.-Petersburg (LE).

On sand steppes.

IA. Mongolia: *Depr. Lakes* (Kytuk river, east of estuary, June 9, 1903—Gr.-Grzh.; Borig-del' sand south-east of Bain-nur lake, July 25, 1945—Yun.).

IIA. Junggar: *Cis-Alt.* (on Kandagatai river, Sept. 14, 1876—Pot.), *Zaisan* ("near Temir-su, in abandoned fields, Aug. 2, 1876, Pot."—Maxim. l.c.).

General distribution: Aralo-Casp., Fore Balkh., Jung.-Tarb.; West. Sib., Nor. Mong.

Note. Specimens, for example from Hangay, are intermediate to *D. ramosissimus* and *D. versicolor*.

5. **D. superbus** L. Fl. Suec. ed. 2 (1755) 146; Ser. in DC. Prodr. 1 (1824) 360; Ledeb. Fl. Ross. 1 (1842) 285; Kar. et Kir. in Bull. Soc. natur. Moscou, 15, 1 (1842) 166; Turcz. in Bull. Soc. natur. Moscou, 15, 3 (1842) 568; Forbes et Hemsl. Index Fl. Sin. 1 (1886) 64; Will. in J. Bot. (London) 34 (1898-1900) 427; Maxim. Enum. pl. Mong. (1899) 83; id. Fl. Tangut. (1899) 81; Kryl. Fl. Zap. Sib. 5 (1931) 1100; Schischk. in Fl. SSSR, 6 (1936) 856; Hao in Bot. Jahrb. 68 (1938) 594; Walker in Contribs. U.S. Nat. Herb. 28, 4 (1941) 613; Grub. Konsp. fl. MNR [Conspectus of Flora of Mongolian People's Republic] (1955) 134; id. Opred. rast. Mong. [Key to Plants of Mongolia] (1982) 105; Ma Yu-chuan in Fl. Intramong. 2 (1978) 189; Claves pl. Xinjiang. 2 (1983) 253; Zhao in Acta sci. natur. univ. Intramong. 20, 1 (1989) 109. —*D. longicalyx* Miq. in J. Bot. Neerl. 1 (1861) 109; Zhao in Acta sci. natur. univ. Intramong. 20, 1 (1989) 109. —*D. hoeltzeri* Winkl. in Regel, Gartenfl. (1882) 1, tab. 1032, fig. 2; Schischk. l.c. 134; Opred. rast. Sr. Azii [Key to Plants of Mid. Asia] 2 (1971) 309; Grub. l.c. 106; Claves pl. Xinjiang. l.c. 252. —Ic.: Fl. Intramong. 2, tab. 100, fig. 6-8.

Described from Lapland. Type in London (Linn.).

In meadows, scrubs, sparse forests and their borders, birch groves.

IA. Mongolia: *Khobd., Mong. Alt., Cent. Khalkha, East. Mong., Depr. Lakes, Alash. Gobi.*
IIA. Junggar: *Tien Shan.*
IIIA. Qinghai: *Nanshan* (between Cheibsen temple and Yuzhno-Tetungsk mountain range, in fields and meadows, July 29; "in Yuzhno-Tetungsk mountain range, 2250 m,

in forest grasslands, Aug. 10, 1880—Przew.; on Itel'-gol river, Aug. 1885, Pot."—Maxim. l.c.).

General distribution: Jung.-Tarb., Tien Shan; Europe, Caucasus, West. Sib., East. Sib., Far East, Nor. Mong., China (Nor., Nor.-West., Cent., East., South-West., South., Hainan), Korean peninsula, Japan.

6. **D. versicolor** Fisch. ex Link, Enum. pl. Horti Berol. 1 (1821) 420; DC. Prodr. 1 (1824) 358; Turcz. in Bull. Soc. natur. Moscou, 15, 3 (1842) 567; Will. in J. Linn. Soc. (London) Bot. 29 (1893) 424; Kryl. Fl. Zap. Sib. 5 (1931) 1103; Schischk. in Fl. SSSR 6 (1936) 824; Grub. Konsp. fl. MNR [Conspectus of Flora of Mongolian People's Republic] (1955) 135; Opred. rast. Sr. Azii [Key to Plants of Mid. Asia] 2 (1971) 303; Grub. Opred. rast. Mong. [Key to Plants of Mongolia] (1982) 106; Claves pl. Xinjiang. 2 (1983) 254; Fl. desert. Sin. 1 (1985) 465. —*D. elatus* Ledeb. Fl. Alt. 2 (1830) 136; id. Fl. Ross. 1 (1842) 280; Schischk. l.c. 833; Claves pl. Xinjiang. l.c. 252. —*D. turkestanicus* Preobr. in Izv. Glavn. bot. sada SSSR, 15 (1915) 366; Schischk. l.c. 827; Claves pl. Xinjiang. l.c. 254; Fl. desert. Sin. 1 (1985) 466. —*D. chinensis* var. *versicolor* (Fisch. ex Link) Y.C. Ma in Fl. Intramong. 2 (1982) 191; Zhao in Acta sci. natur. univ. Intramong. 20, 1 (1989) 110. —*D. subulifolius* Kitag. in Rep. first sci. exp. Manch. sect. 4, 2 (1935) 16. —**Ic.:** Fl. Intramong. 2, tab. 101, fig. 1-3; Fl. Kazakhst. 3, Plate 45, fig. 2, Plate 46, fig. 6; Fl. desert. Sin. 1, tab. 170, fig. 3-5.

Described from East. Kazakhstan. Type in St.-Petersburg (LE).

In steppes and meadows, rubbly slopes, on streams, sparse forests, scrubs.

IA. Mongolia: *Mong. Alt., Cent. Khalkha, East. Mong., Depr. Lakes, Gobi-Alt.*

IIA. Junggar: *Cis-Alt.* (20 km nor.-west of Shara-Sume, July 17, 1959—Yun.), *Tien Shan* (south-east. bank of Sairam, 1878; Kyzemchek, Sairam, July 29, 1878; Sairam, June 20; Talki gorge, July 18; Kul'dzha, July 1877; Borgaty (Kash), July 4; Arystyn, July 11, 1879—A. Reg.; westward of Sairam lake, July 28, 1878; Urtak-sary, July 20; near Sairam lake, July 23, 1878—Fet.; Ketmen' mountain range, nor. slope, 8-10 km south of Sarbushin on Ili—Kzyl-kure road, Aug. 23; Shuvutin-daba pass, north of Sairam-nor lake, Aug. 18—Yun. and I.-F. Yuan'), *Zaisan* (Khobuk river valley, July 20, 1914—Sap.; Chern. Irtysh river, right bank near Burchum estuary, tugai, June 14, 1914—Schischk.).

General distribution: Aralo-Casp., Fore Balkh., Jung.-Tarb., Tien Shan; Europe, West. Sib., Nor. Mong., Far East.

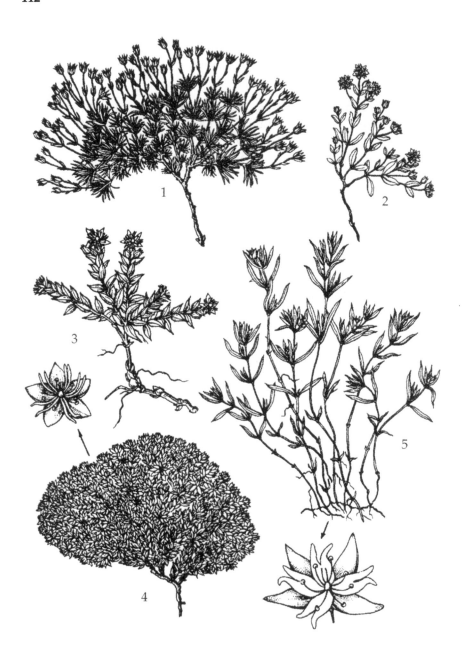

Plate I

1 — *Stellaria cherleriae* (Fisch ex Ser.) Will.; 2 — *S. winkleri* (Briq.) Schischk.;
3 — *S. arenaria* Maxim.; 4 — *S. maximowiczii* Yu. Kozhevn.; 5 — *S. alexeenkoana* Schischk.

Plate II

1 — *Stellaria depressa* Schmid; 2 — *S. divnogorskajae* Yu. Kozhevn.; 3 — *S. dichotoma* L.;
4 — *S. gypsophilloides* Fenzl; 5 — *S. amblyosepala* Schrenk; 6 — *Melandrium alaschanicum*
(Maxim.) Y.Z. Zhao

114

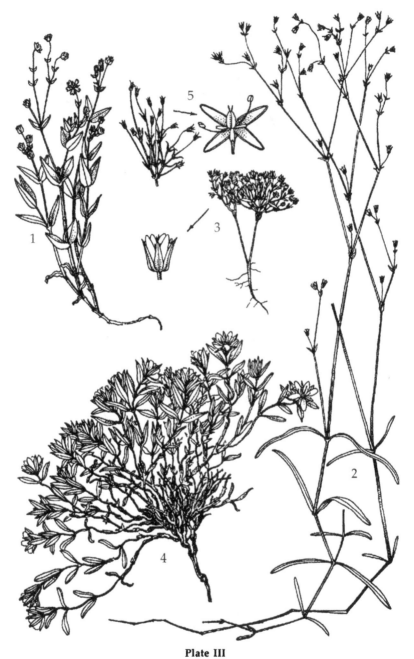

Plate III

1 — *Stellaria merzbacheri* Yu. Kozhevn.; 2 — *S. viridescens* (Maxim.) Yu. Kozhevn.;
3 — *Sagina karakorensis* (Schmid) Yu. Kozhevn.; 4 — *Cerastium cerastoides* var. *foliosum*
Yu. Kozhevn.; 5 — *Arenaria littledalei* Hemsl.

Plate IV

1 — *Minuartia kryloviana* Schischk.; 2 — *Melandrium glandulosum* (Maxim.) Will.;
3—*Lepyrodiclis quadridentata* Maxim.; 4—*Stellaria pusilla* Schmid; 5—*S. depressa* Schmid;
6 — *Sagina karakorensis* (Schmid) Yu. Kozhevn.

Plate V
1 — *Cerastium alpinum* L.; 2 — *C. pumilum* Curt.; 3 — *Acanthophyllum pungens* (Bge.)
Boiss.; 4, 5 — *Silene viscosa* (L.) Pers.

Plate VI
1 — *Silene alexandrae* Keller; 2 — *S. odoratissima* Bge.; 3 — *S. subcretacea* Will.; 4 — *Stellaria gyangtsensis* Will.

118

Plate VII

1 — *Arenaria edgeworthiana* Majumdar; 2 — *A. melanandra* (Maxim.) Mattf.; 3 — *A. melandryoides* Edgew.; 4 — *A. caespitosa* (Camb.) Yu. Kozhevn.; 5 — *A. przewalskii* Maxim.; 6 — *A. glanduligera* Edgew.

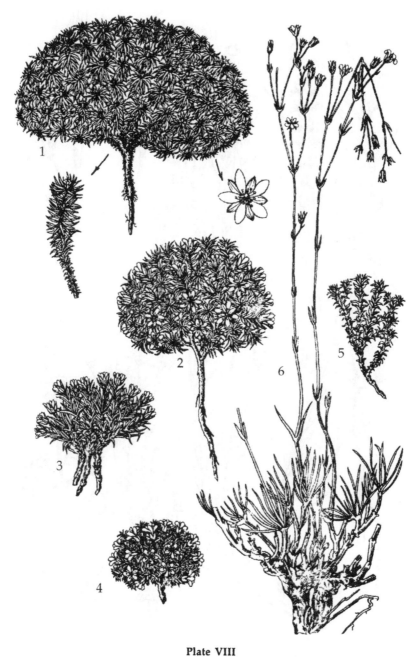

Plate VIII
1 — *Arenaria polytrichoides* Edgew.; 2 — *A. roborowskii* Maxim.; 3 — *A. kansuensis*
Maxim.; 4 — *A. pulvinata* Edgew.; 5 — *A. densissima* Wall. ex Edgew.; 6 — *A. pentandra*
Maxim.

Plate IX

1 — *Silene maximowicziana* Yu. Kohevn.; 2 — *S. lithophila* Kar. et Kir.; 3 — *S. nana* Kar. et
Kir.; 4 — *S. holopetala* Bge.; 5 — *Gymnocarpos przewalskii* Bge. (5a — single flower)

Plate X

1 — *Gypsophila microphylla* (Schrenk) Fenzl; 2 — *G. trichotoma* Wend.; 3 — *G. cephalotes* (Schrenk) Will.; 4 — *G. capituliflora* Rupr.; 5 — *Dianthus barbatus* L.; 6 — *D. crinitus* Smith.

122

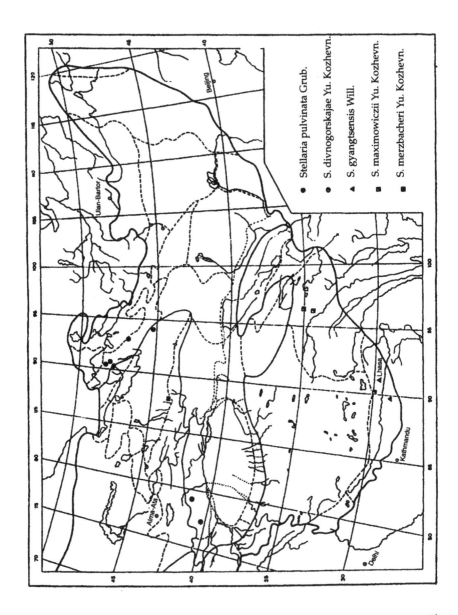

Map 1

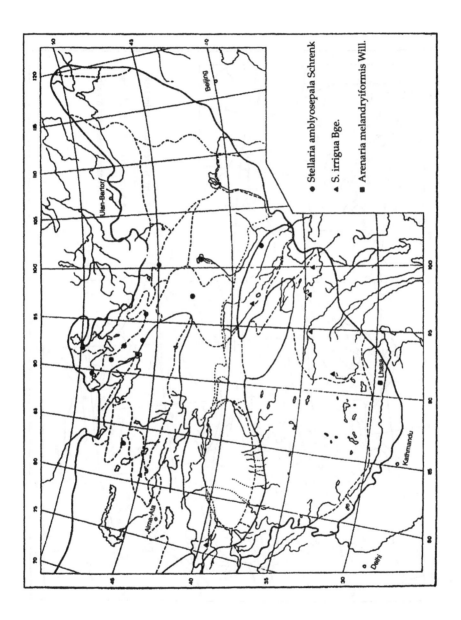

Map 2

Legend:
- ● Stellaria amblyosepala Schrenk
- ▲ S. irrigua Bge.
- ■ Arenaria melandryiformis Will.

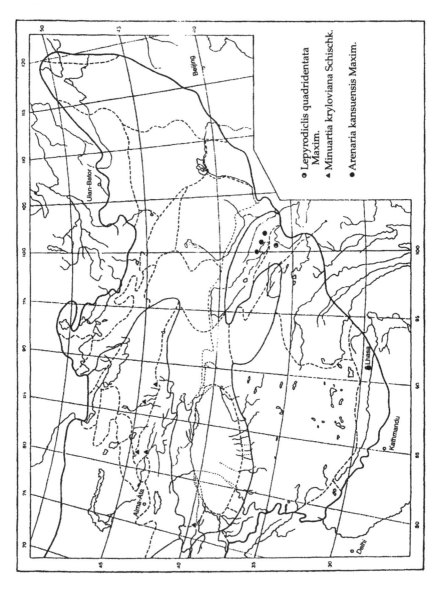

● Lepyrodiclis quadridentata Maxim.

▲ Minuartia kryloviana Schischk.

● Arenaria kansuensis Maxim.

Map 3

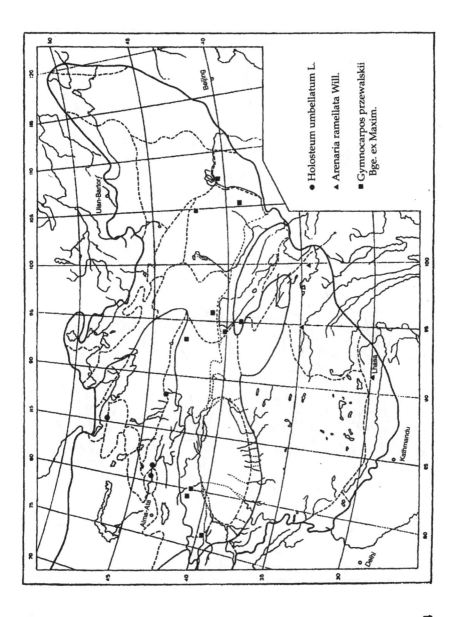

Map 4

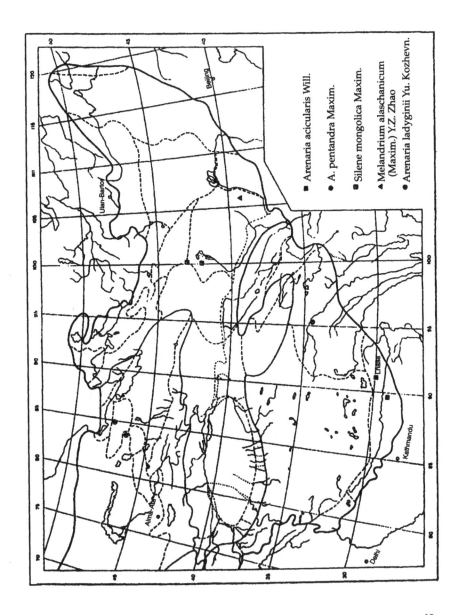

■ Arenaria acicularis Will.

● A. pentandra Maxim.

▣ Silene mongolica Maxim.

▲ Melandrium alaschanicum (Maxim.) Y.Z. Zhao

● Arenaria ladyginii Yu. Kozhevn.

Map 5

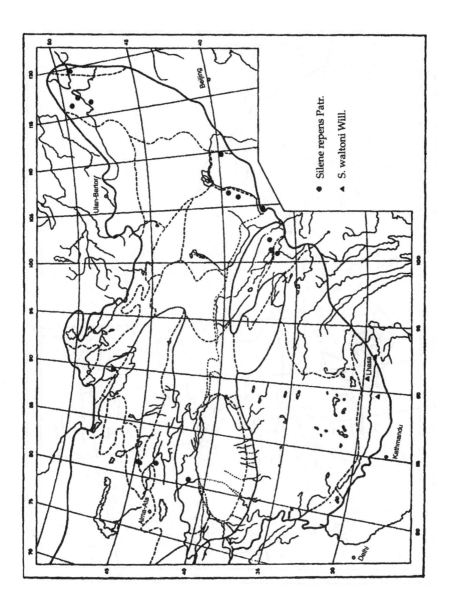

Map 6

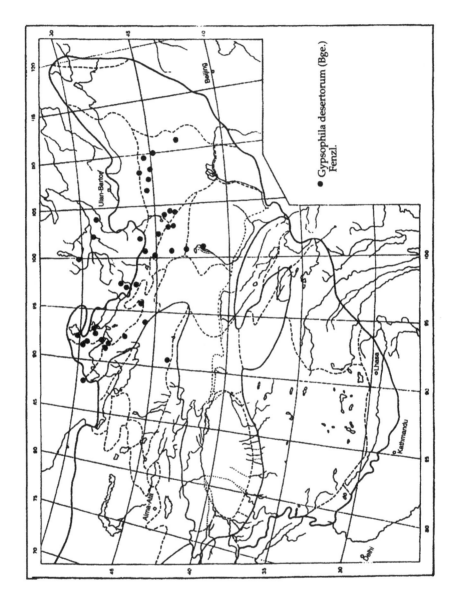

• Gypsophila desertorum (Bge.) Fenzl.

Map 7

INDEX OF LATIN NAMES OF PLANT

INDEX OF PLANT DRAWINGS

INDEX OF PLANT DISTRIBUTION RANGES

For Product Safety Concerns and Information please contact our EU
representative GPSR@taylorandfrancis.com Taylor & Francis Verlag GmbH,
Kaufingerstraße 24, 80331 München, Germany

Printed and bound by CPI Group (UK) Ltd, Croydon, CR0 4YY

01/05/2025

01858551-0001